Lecture Notes in Physics

For information about Vols. 1–131, please contact your bookseller or Springer-Verlag.

Lecture Notes in Physics

Edited by H. Araki, Kyoto, J. Ehlers, München, K. Hepp, Zürich
R. Kippenhahn, München, H. A. Weidenmüller, Heidelberg
and J. Zittartz, Köln

196

WOPPLOT 83
Parallel Processing: Logic, Organization, and Technology

Proceedings of a Workshop Held at the
Federal Armed Forces University Munich (HSBw M)
Neubiberg, Bavaria, Germany, June 27–29, 1983

Edited by J. Becker and I. Eisele

Springer-Verlag
Berlin Heidelberg GmbH 1984

Editors

Jörg D. Becker
Ignaz Eisele
Institute of Physics, Faculty of Electrical Engineering
Federal Armed Forces University Munich
D-8014 Neubiberg

ISBN 978-3-540-12917-2   ISBN 978-3-540-38803-6 (eBook)
DOI 10.1007/978-3-540-38803-6

Originally published by Springer-Verlag Berlin Heidelberg New York Tokyo in 1984

2153/3140-543210

Preface

WOPPLOT 83 - a Workshop on Parallel Processing: Logic, Organization and Technology - was held on the campus of the Federal Armed Forces University Munich.

Wopplot was meant to yield an opportunity to look and talk across the fences of the various and all too separate fields of research, sounding possible lines of development in parallel processing. The workshop was concentrating on

- physical and technological foundations,
- architectures and algorithms, and
- cybernetic aspects of organization.

The interdisciplinary character of the workshop also led to the inclusion of several talks which do not have an immediate impact on parallel processing but which might have some significance in the future.

The "Logic" called for in our invitation does not show up appropriately in these proceedings. We decided, however to leave the name of the meeting as it is, wishing WOPPLOT to remain an active nucleation site.

IV

It is a pleasure to thank all twenty-three participants from five
European countries and from Japan for their coming and for the live-
ly discussions. In particular the editors would like to thank all
the speakers for their carefully prepared talks and for their manus-
cripts.

For all those who made the technicalities of the meeting run so
smoothly we wish to acknowledge Mrs. E. Göppel, Mr. B. von Hornstein,
and Mr. H. Pechtold.

Finally, particular gratitude for cooperation and financial support
is due to our sponsors:

Siemens AG, Munich
Messerschmitt-Boelkow-Blohm GmbH, Ottobrunn
Freundeskreis der Hochschule der Bundeswehr München, Neubiberg

Neubiberg, December 1983

J. Becker   I. Eisele

C O N T E N T S

Preface

CYBERNETIC PRINCIPLES OF ORGANIZATION

F. Vester

Federal Armed Forces University
Institute for Independence of Technical and Social Change
D-8014 Neubiberg, Federal Republic of Germany

I.   INTRODUCTION

In our common language we use quite a lot of associations with other
informatory perceptions than speech: we describe pictures, movements,
give examples, produce feelings. In other words, we try to connect
the speech center in our brain with other cerebral areas like the
visual cortex, the cerebellum with its haptic and motive functions
or the hypothalamus and its emotional interpretations. In our
scientific language we rather try to stick to abstract words, anxious
not to touch those other informational channels.

These channels, however, are as accurately working, as uncorruptible
- or corruptible - grey cells as those which work in the causal-logical
regions of our brain. They have even distinct advantages: visual im-
pressions, for instance, are much better in parallel processing. They
enter by simultaneously registering millions of bytes recognizing
pattern in a fraction of a second: a far greater performance than the
linear, sequential processing of words.

As in this paper I shall stress the necessity to understand our
environment not as a heterogeneous list of separate items but as a
pattern, a network of relations, I think it is valuable to use as
manny slides as possible - view taken from reality - to accompany my
talk. This will not only help remembering by a better associative
storage but also help to build up a parallel understanding of what
I say, with additional parts of our brain.

Being a biologist originally, I had to deal with organisms, that is
with open complex systems, from the beginning. And the larger these
systems became - cells, organs, individuals, ecosystems - the more the
common laws of their organization became clear to me. The ecosystem
approach, therefore, seems applicable to all levels of life. Two of
them I will touch in my talk: the level of the brain, as an organ
that reflects already in its hardware the environment it was born in
(by anatomic storage of the first perceptions - until about three
months after birth) and how we can train it (and help it by computer-
ized simulation) in the skill of 'pattern recognition'. Secondly, the

level of ecosystems, understood as 'organisms' obeing some fundamental systemic laws of survival. Both acting on each other: the brain by using its structure to understand and handle the engrammed environment, and the environment by imprinting directly or indirectly (via science) our brain. This latter in very different degrees of understanding the pattern of that environment.

In so far as we act according to this understanding, our environment is treated differently and so are the ways of our economic strategy and their effects. Let me therefore first review the effects of our classical "linear thinking" and of the corresponding planning strategy.

II.   THE ISOLATED SYSTEMS APPROACH AND THE CRISIS OF CIVILIZATION

INTERFERENCE WITH ENVIRONMENT

Throughout the course of human history,in our believe that the buffer capacity of nature is unlimited, its realm of resources is endless and that what is technically possible is also achievable - with these illusions we have rather carelessly interfered with a - hithereto perfectly - functioning and evolving system, our biosphere, in that we have gradually imposed more and more artificial systems upon it:

> Factories, power stations, and large-scale farmings.

> Housing estates, reservoirs, road networks, bridges,and ports.

> Natural landscapes, such as those in North America, often had to give way to sprawling urban systems within only a few decades.

We grafted all these systems - be it traffic networks or the Aswan High Dam with all its implications - on our biosphere in the assumption that their interaction, their communication, would regulate itself and remain benificial to men who invented them.

REPAIR SERVICE BEHAVIOUR

If anything went wrong, we thought, all that would be needed to repair matters would be the application of sufficient technology and of sufficient energy to the problem. Thus we thought we would be able to make good any deficiencies in the purity of the air, the fertility of the soil, energy supplies, the natural water balance or in the health sector. Confident, that there nothing we could not repair, we neither cared whether these artificial systems would be viable if left to themselves nor whether their interactions could manage the load, nor whether they could be linked with others to form a properly functioning unit.

And still, day by day, we initiate further development projects and set them between existing systems, without even knowing that we are dealing with systems, without knowing that anything such as laws of systems behaviour exist, and rules that determine the chances of survival of a system.

On this basis we also exert a strange behaviour towards problems. In our times of critical constellations and crisis one can observe a set-back of constructive tasks while the attention of our decision-makers is directed more and more towards obvious defects and nuisances and their activity towards their repair, or elimination. A defect in an open system however is something very different from a defect in a machine. There one can exchange parts or glue something together and the machine is working again. In an open system any repair which is concentrated on the defect part only, will have side-effects, thus forcing us very soon to repair the repair and to repair the repair of the repair and so on.

The result is, that we slided inexorably into our actual repair-service behaviour which creates only new problems, is more and more expensive, makes reasonable planning impossible and leaves us lagging constantly behind general development. What we need is not to repair the defects nor to forecast it and when they will occur (another illusion!) but to create constellations which give those defects less chance to destabilize our system in consideration.

As long as we do not understand this, we are faced with the fact that many of our interferences with the environment which were designed for quick profits will first affect the environment, then the quality of our own lives and, as a final effect, will become great economical problems too - especially in developing countries. There are enough examples of this.

SIX ASPECTS OF OUR INABILITY TO DEAL WITH COMPLEX SYSTEMS

<u>Firstly</u>: energy-squandering consumer goods and the growth of private transportation make us and our sub-systems more and more dependent on energy. The same occurred through the increasing use of energy-intensive materials and production processes. Such a course of development in energy use is no progress, because it threatens the stability of the system.

Therefore, in living nature any step forward in evolution was generally accompanied by an amelioration in energy efficiency, that is: towards less energy consumption per biomass-unit. Compared with that, our way in the technology of the last hundred years was a clear step backward in evolution.

Secondly: we are making ourselves dependent on unrecoverable raw materials in much the same way - and throw them away after less and less time of use. Nature, on the other hand, with her refined technologies has been processing hundreds of billions of tons of carbon and oxygen and also thousands of millions of tons of heavy and light metals such as iron, magnesium, potassium and calcium, year by year, century by century since her birth without ever having raw material or waste disposal problems. Thanks to her clever recycling and symbiotic techniques, every waste product immediately becomes a new raw material.

Thirdly: we deliberately change landscapes, carry out forced cultivation in a completely unecological manner, clear more and more land as the soil loses its vitality. This applies for both forests and arable land. Apart from this, with our monocultures, mass production of livestock and corresponding high energy consumption and transport, we cultivate forage plants on a profiligate scale in the Third World, something that can be held partly responsible for the shortage of food in developing countries.

In doing so, we use huge amounts of energy-intensive fertilizers, destroying profitable equilibria such as the surface water's power of self-regeneration, we destroy birds and insects, although the work they do as a properly functioning and balanced system brings us profits that count in billions.

Fourthly: similar effects result from our thoughtless building of entire new suburbs, often on the basis of only short-lived criteria. What happens in so-called "urbanization projects" is, that one is left with social and financial costs that neither the citizens nor the community can bear any longer.

Fifthly: Not only do we fail to see our environment, our cities and landscapes, as parts of a system and therefore ignore important system laws in our planning, we also fail to see ourselves as part of the system. One could therefore say that the fifth main aspect of our non-systemic strategy is, that also medicine and psychology are drifting into an expensive "repair service"-type of behaviour instead of placing them-

selves at the service of the only thing that is profitable, the pre-
vention of disease.

Here once again, we find interferences in individual sectors - with
much the same sort of results as we have already seen. The medical
profession first repairs, and then carries out further repairs to
repair the repairs.

THE PRESSURE OF DENSITY STRESS

Medical statistics and that mental diseases, growing sterility, etc.
shows, that a sort of self-regulation can be observed for the human
population. A negative feedback in the form of "density stress" seems
to work, a mechanism, that ensures that a population which is growing
to fast, drastically reduces its own number to achieve a lower density
that again will allow survival. This mechanism seems to me quite of
interest since, once begun to take effect, the density stress mechanism
leaves only two possibility open: it either causes stress-induced
agressions, diseases, sterility and reduced brood-care instinct (via
a psychosomatic mechanism) all of which lead to the reduction of a
large part of the population and thus back to the earlier density. Or
- this is the evolutionary way out, it forces populations to change
their behaviour, to progress to a higher organizational form that
allows them to survive, even with higher population density.

The question of rearrangement in organization and communication
appropriate to our high level of density and interdependence - which
may be tenfold to medieval times - is therefore vital. As vital as
it was a few thousand years ago before the gatherer-and-hunter-society
changed to the economy of planters and heardsmen (a jump from 400.000
to 14 million people on earth). But instead of learning how to deal
with and how to organize complex systems we constantly interfered with
them without recognizing them as systems and we continue to act as we
were dealing with a heterogeneous quantity of individual things, while
many parts of our world that previously were independent indeed, became systems or parts of systems just by their increasing density.

One of the main causes of this dilemma is our lack of knowledge of the
interrelationships of the cybernetic laws of systems, of the rules of
survival and the believe in the possibility of forecasting and de-
terministically constructing our future. All this, in turn, results
from the very nature of our non-systemic education.

III. DYNAMIC NETS AND INTERACTIVE SCENARIOS

The crux of the matter is that we (with the analytical way of thinking
we learned) concern ourselves in detail with individual mechanisms and
individual structures but practically never with the dynamic network
between them. However, the real world in which we live is not what
they taugh us at school and university: a hotchpotch of separate sectors
such as agriculture, transport engineering, chemistry, geography,
commercial management, building industry and waste disposal - all
clearly arranged in departments and sectors and thus reduced to frag-
ments of what is really a cross-linked system that behaves in accordance with cybernetic laws. It is the net we do not realize and consider. The actual systems character is beyond our comprehension. It
breaks up the scientific fields, one cannot even assign it to a
faculty. So it remains unconsidered.

THE CASE OF REGIONAL DEVELOPMENT

Let us take an example of regional development. We know the things we
are dealing with - roads, houses, factories, raw materials, forests
and naturally also ourselves - only as roads, houses, factories, raw
materials, forests and people, and this is how we treat them. We do
not know them in their cybernetic function, which means their different
roles in the open cross-linked system which represents the region in
question.

With their real interdependences, however, the things we know (by one
definite name only) play the roles of controllers, control elements,
probes, buffers, limit values or replenishment values - always different
in different cases. These roles we ignore completely. How, therefore,
can we know of the cybernetics character of a system that is made up
of such things: its tendency to stabilize itself, its susceptibility
to disturbances, its flow equilibrium, its external and internal dependencies, the interlinking of its feedback cycles or its diversity,
its variety? All that can never be drawn from the isolated elements,
but only from what happens between them.

THE OUTCOMES OF LINEAR PLANNING

The fact that we have shown scarcely any interest in the interactions
of complex systems has a second consequence. It is certainly one of
the reasons why cybernetic technologies that have long been possible
but, however, require, an appreciation of cross-linkages, are still in
their infancy. This is why we have scarcely any symbioses, scarcely

any recycling, energy chains and multiple usages like in the case
of combined composting, biogas, water purification and heat exchange
unit, nor other forms of what is an elegant small-scale and therefore
all the more efficient biotechnology like photosynthetic plants. Nor
do we find interlocking factories of energy- or matter-compensating
branches with their exchange flows of raw material, waste and energy -
such as would correspond to a suitable eco-system of industrial settle-
ment. Not recognizing the system, we design them separately - each has
its own input of matter and energy and its own output of waste - and
thus we don' t know where and how we destroy or ignore profiting cycles
or self-controlling feedback systems, where and why we suddenly come
up against unexpected barriers, or why our planning fails.

The results are <u>unexpected problems</u> in accordance with the previously
mentioned examples. Secondly, we turn to "<u>solutions</u>" that never really
solve anything: support from the state, that only perpetuate obsolete
forms of industrial or economic structure, or enhancing growth that
causes entire regions or branches of industry to practically collapse
like the shut-down steel plants on the Saar. But the same is true for
the Peruvian fisheries, the collapse of the Laker Airlines or the
Wienerwald imperium or the Crysler collapse in the U.S.A. or the breakdown of whole countries like in South America - to give only a few
widely-differing examples. Thirdly, we continue to use <u>technologies</u>,
although they are patently absurd - the problematic nuclear energy
and its growing number or outdated power-plants, while costs will rise,
or the Concorde adventure in France and its obsolete services - or the
great number of useless supertankers built in an euphoric boom - and
now ruining one shipping company after the other. And forthly: we
develop forms of <u>organization</u> that can never stand the acid test of
reality - incomprehensible bureaucracy, the increasing centralization
of supplies, and monostructures in agriculture and industry with rising
operation costs and rising dependencies. So far the main reactions of
the system upon linear planning.

NEGATIVE FEEDBACK - ARTIFICIALLY BLOCKED

The reasons why a systemic approach must lead to a better understanding
of reality are easily found in the cybernetics behind it. We saw that
interventions in an open dynamic system always have very complex re-
percussions. Only a small percentage of these are expressed in a
direct cause-and-effect relationship and practically none in a
straight line.

A simple example: the attraction for touristic activity rises with the accessibility of a certain landscape.

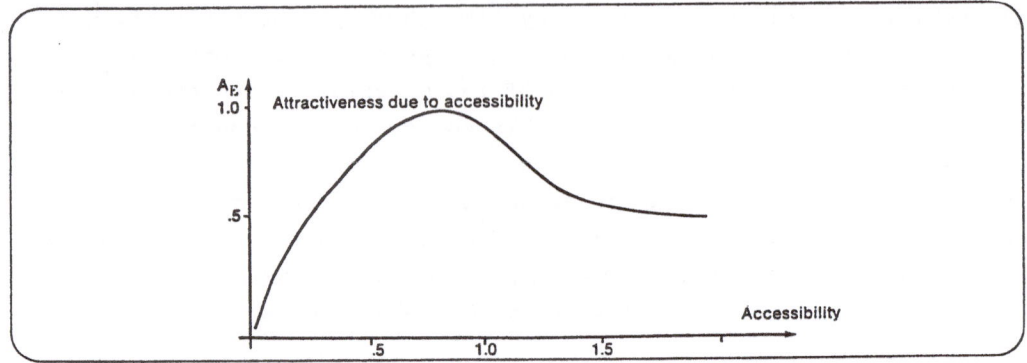

Fig. 1.   Table function: Attractiveness for tourism
          due to accessibility of landscape.

Already before an optimal infrastructure is reached, the curve begins to incline because of the draw-backs from rising traffic and deterioration of nature. Therefore it would be desastrous to extrapolate the linear trend, corresponding to this short part of the curve.

Like in this case, many observed developments of data, serving for marketing strategies, are only parts of much more complex curves or even networks of curves. At first sight, however, these relationships appear to take a linear course and to grow proportionally. But because of their involvement of the system as a whole, they soon assume threshold and limiting values, not at first perceived, which suddenly distort their uniform development.

In many cases this ressembles the case of the bow and the arrow. Before a certain threshold-value you can move the arrow forth and back without anything happens. This is the first stage. Above this threshold, the more you pull the string, the further flies the arrow. An almost proportional relationship. This is the second stage. If you pull even stronger you eventually reach a third stage: the bow brakes and the arrow doesn't fly at all anymore.

In Nature a system stabilizes itself by negative feedback, before a dangerous limiting or boundary value is reached. We frequently, however, remove this self-regulating mechanism by additional interventions, an artificially induced boom, the introduction of subsidies, the

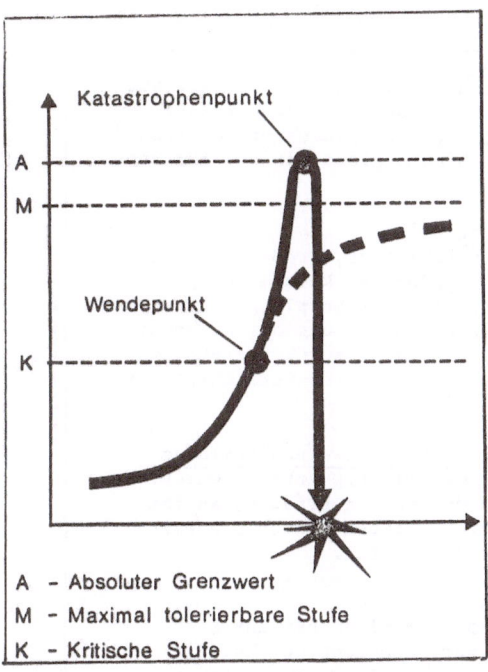

Fig. 2. A system deviating from the S-shaped logistic curve of growth will eventually reach absolute limit values causing its breakdown.

pumping in of extra energy, redoubling the exploitation of natural resources and thus overstep the limiting values, with the result that we soon come up against higher limiting values which then often lead to the collaps of the system concerned.

Take the case of artificially protected elephants in a natural reserve area. Their growing number destroyed their nutrional base, the acacia trees. And with the last leaf eaten the whole heard died by starvation. With less protection (= less population growth they would have had no problem).

In all these cases, we are mislead by simple causal chains, we fore-cast by extrapolation, take lines of thought that concern themselves with individual problems only. In the past, in an area of growth, where systems behave temporarily like a machine, we managed for quite a while this way. But to be able to live with the cross-linkages of our actual situation and to make use of them in the same way as Nature does, we need a completely new approach that will allow us to under-stand such systems. One could call it "biocybernetic thinking", a thought pattern that would automatically eliminate certain fundamental mistakes made in planning. Mistakes that have been listed in six groups by the system-psychologist Dietrich Dörner[1] of the University of Bamberg (see the following table).

The Six Errors in Dealing with Problems in Complex Systems (Dörner[1])

Erster Fehler: Mangelhafte Zielbeschreibung
Das System wird abgetastet, bis ein Mißstand gefunden wird .Dieser wurde beseitigt.Dann wurde der nächste Mißstand gesucht und dann wieder vielleicht eine Folge der ersten Aktion korrigiert. Man nennt so etwas Reparaturdienstverhalten.Die Planung geschah ohne große Linie , ähnlich wie bei einem Anfänger im Schachspiel.

Zweiter Fehler: Unvernetzte Situationsanalyse
Einige Versuchspersonen waren immer damit beschäftigt, große Datenmengen zu sammeln, die zwar enorme Listen ergaben, jedoch zu keinem Gefüge führten.Wegen fehlender Ordnungsprinzipien, also Rückkopplungskreisen, Grenzwerten usw. gelingt dabei natürlich keine Auswertung der Datenmassen. Denn auf die Erfassung des dynamischen Charakters des Systems, wie zum Beispiel auf seinen geschichtlichen Werdegang, wurde verzichtet. Die Dynamik des Systems blieb unerkannt.

Dritter Fehler: Einseitige, zum Teil irreversible Schwerpunktbildung
Man versteifte sich auf einen Schwerpunkt, der richtig erkannt wurde. Er wurde zum Favoriten. Man biß sich aufgrund des ersten Erfolges an ihm fest und lehnte andere Aufgaben ab. Hierdurch blieben jedoch schwerstwiegende Konsequenzen in anderen Bereichen unbeachtet.

Vierter Fehler: Unbeachtete Nebenwirkungen
Im linear-kausalen Denken befangen, geht man vor allem bei der Suche nach geeigneten Maßnahmen - mit denen man zur Besserung der Lage in das System eingreifen kann - ohne Nebenwirkungsanalyse vor - oft auch dann noch, wenn man das System selbst als vernetztes Gefüge erkannt hat. Man unternimmt sozusagen keinen Policy-Test.

Fünfter Fehler: Tendenz zur Übersteuerung
Eine häufige Vorgehensweise, die Dörner beobachtete, war folgende: man ging zunächst sehr zögernd und mit kleinen Eingriffen heran; wenn sich dann im System nichts tat, war die nächste Stufe ein kräftiges Eingreifen, um dann bei den ersten unerwarteten Rückwirkungen - durch Zeitverzögerung hatten sich die kleinen Schritte vielleicht zunächst akkummuliert - wieder komplett zu bremsen.

Sechster Fehler:Tendenz zum autoritären Verhalten
Die Macht das System verändern zu dürfen, und der Glaube, es durchschaut zu haben, führt zum Diktatorverhalten, welches für komplexe Systeme völlig ungeeigner ist.Für diese ist ein anschmiegsames Verhalten, das nicht gegen den Strom, sondern mit dem Strom schwimmend verändert, am wirkungsvollsten. Hier spielt übrigens unsere Grundregelvom Jiu-Jitsu mit hinein, also möglichst Nutzung vorhandener Kräfte durch Umlenkung, statt das üblich Durchboxen gegen dieselben.

These six errors explain the desastrous results of the famous "Tanaland"-experiment of Dörner[1]. A cybernetics expert of the BASF (E. Schmäing) called the experiment an example for the fact that particularly those crisis-management staffs which operate logically are overdemanded in their attempts to improve the critical situation of a network-system. Such a system, because of its unnoticed interconnections, apparently behaves counter-intuitively, i.e. the measures applied to it do not result in what one "logically" would expect of them.

THE PICTURE OF ABRAHAM LINCOLN

Now, what means cybernetic thinking? It essentially relies upon
pattern recognition, the interpretation of the main features of a
system, something that computers are known to have great difficulties
with. What about our brain? I think, it can indeed change between two
different ways of perceiving reality. The difference between the mono-
causal form of recognition (which reviews individual data) and pattern
recognition (which reviews the character of a system) can be demonstrated
by an example.

When we look at the squares of differing brightness in the left picture
from a closer distance, we have difficulty in recognizing what it represents. As soon as we alter the focus, or by squinting a little or
taking off our glasses (the right picture has been taken that way) we
immediately recognize the features of U.S. President Abraham Lincoln.

Suddenly we can understand what this is trying to tell us. In the
latter case, the groups of brain neurons that go into action are
completely different from those that are working when we concentrate
on the details, the small squares. Our brain is able to recognize
the whole in spite of missing parts. In removing the detail by
fuzziness, the relations between the squares become predominant and

reveal the character of the system. This is pattern recognition, the ability to recognize the system's interrelationships.

This example tells us a lot about our topic of thinking in systems. It tells us that a detailed study of the individual squares in our photo may reveal only a certain kind of information: the exact gray values, a table of the edge lengths, or the percentage of squares along a scale of different brightness - that is the way doctor-theses are carried out - but it will never allow us to recognize (like the fuzzy pattern) that it is a portrait of Abraham Lincoln. Studying the single squares it is therefore the wrong scientific method for systems and does not become anymore correct when we do so with scrupulous accuracy. In trying to recognize a system, the immediately apparent details are of no help, useful as they may be in selecting the operators of a strategy later. On the contrary: the foggier the focussing of details is, the clearer the relationships between them become, the easier it is for us to say what the picture as a whole represents - so to say, the system and its behaviour.

OPEN SYSTEMS: COMMUNICATION WITH ENVIRONMENT

To understand such a system we therefore have two ways in science. We can follow the same course as employed for all the other things we study, by isolating it and investigating it as a self-contained whole. This mechanistic approach, although applicable in so many other cases, will however fail dismally when we apply it to a complex dynamic system. Treating what is really an open system, as if it were closed, ignores one of its most important characterists:its complex behaviour in relationship with its environment. This form of open organization can only be recognized, when one has a knowledge of the dynamics of its internal and external channels of communication, regardless of whether one is dealing with a single human cell, for which over 10.000 internal and external channels of communication are plotted on any biochemical metabolic chart, or with a system we call an urban district.

The knowledge of this primary structure and its dynamics therefore is important. But is is only the first level of understanding a system the ecosystems-approach with its aim of survival exceeds this level of pure simulation as,for instance, in the method of systems dynamics used for the "world model" or the "limits of growth. As Wilson and Clarke[2] of the University of Leeds point out: the ecosystems-approach enables the decision-maker to enact different roles, and by being interactive, a planning model there upon avoids the pitfalls of those of Meadows and Forrester, where with all the causative links contained within the structure and the initial values of the variables and parameters set, the model generates a future without any further action of the modeller ..." - it works like a machine (this, of course, only when misused as a prognostic instrument).

By interaction with the user, however, the simulation serves as a pattern to be interpreted cybernetically. This interpretation is the second level, above simulation. To this, Probst and Malik[3] (Directors of the "Betriebswirtschaftliche Institut" and the "Management Center St. Gallen") wrote: "The ecosystems research, that investigates the genesis, structure and dynamics of causal-loop-diagrams (Wirkungsgefüge) is for the future top-management probably more important then national economy.

NATURE AS A GUIDANCE

In order to gain a deeper understanding of the qualitative meaning of systems pattern and how decision-makers can influence the course of events, the requisits to survival and other basic principles have been worked out on a third level: that of evaluation. Here we need more than just interpretation, we need judgement, we need to appeal to a higher court. Where to find this authority? Since the problem is survival, I don't know the better one than the one system, which has proven that it can survive for billions of years and through the most unbelievable external attacks: Nature.

Her basic principles can be expressed in eight rules, taken by analogy from the organization of ecosystems. Eight rules, that I first have worked out in our paper "Urban Systems in Crisis"[4] which now makes part of the policy statement of UNESCO's "Men and the Biosphere"-Program, called MAB, and of other practical uses - from new controlling methods up to a new approach to cybernetic management.

In the following my Institute was engaged in a further investigation on regional planning within the MAB-Program to develop an appropriate instrumentarium along these lines, which enables the planner to understand the socio-economic-ecological environment as a biocybernetic system. This has been published in English and German under the title "Sensitivity Model"[5].

By a new kind of device on all three levels: simulation, interpretation, and evaluation, it helps him not to predict the time or place of an event, nor the selling figure of a product in 1985, but it helps him to obtain the badly needed political and material support for decisions, out of the systems behaviour. Decisions, appropriate to improve the future ability of the system to survive and to evolve. Thus, a practical neutral instrumentarium was created applicable to any geographic region as well as to any economic complex such a firm or even a single building like a recreation center.

THE SENSITIVITY-PROCEDURE

Let me point out just a few criteria typical for this new procedure:

1. From the level of data-collecting and data-use, concerning the choice of variables.

2. From the level of cybernetic interpretation concerning the pattern of interaction.

3. From the level of evaluation concerning the eight bio-cybernetic rules of survival.

I begin with the data. One of the main criteria of our approach is, that the respective model deliver a fairly appropriate description of the systems behaviour even with a small number of data, with fuzzy or estimated data, lacking data, as long as the set of these data satisfies the cybernetic criteria of a systemic pattern of interactions.

We found out that - like in the case of the pattern of Lincoln - this is indeed possible as long as a network of interrelations between some representative components of the system can be established. Like in a hologram with damaged parts of lacking data it will always show the complete and correct picture, only a little fuzzier but not wrong. The way it is coded by interlocking networks will prevent this. (If you brake the slide - on the contrary the picture is destroyed -).

This principle was confirmed in earlier investigations. For instance in analysing an ecosystem of the lower Inn-River in Bavaria using only the population dynamics of some water-bird species. Here one quotient could serve as an indicator-variable containing implicitly many other features of the system.

The reason lies in the fact that, different from non-systemic deterministic models, the interpretation is taken from the interdependencies between the variables and not from variables themselves. Let me

just mention some of the simple tools used. The first step of selecting
the right variables (relevant for the system) can be carried out with a
simple "paper computer" as described in our sensitivity study[5]: a matrix
helping to reduce the set of variables without loosing the relevance
for the system. One of the criteria is that these 8 speres of live
(see table) are always contained in the variable set; another that all
three entities of "being", i.e. energy, matter and information be represented by the variables of these realms, and a third that both:
structure and flow are considered as well.

The Eight Spheres of Life to be Considered

1	Economy (industry, agriculture and forestry, raw materials and energy, services, capital, workplaces)
2	Population (birth and death rates, structure, dynamics and migration, manpower)
3	Land use (fallow land, agriculture and forestry areas, marsh areas, special biotopes, settlements, trade, industry, traffic areas)
4	Human ecology (quality of life, well-being, self-realisation, communal life, security, welfare, education, information)
5	Natural balance (air/water/soil/living world, ecology, output)
6	Infrastructure (traffic, tourism, communications, media, supplies, waste disposal)
7	Community and public sector (regional and communal budget, taxes, public measures and services, decrees)
8	Fringe conditions (basic data and constants of the system including climate, geology, orography, total area etc., which act more or less as a lattice of conditions)

On the second level, that of interpretation, again a simple "paper
computer" serves as an intellectual help to find out the rôle of the
systems components,their meaning in the interlocking pattern and its
dynamics. The validity of this approach is independent of the kind of
the system. It works as well with the structure of the Federal Postal
Administration, as we have tried it in an exercise, or with the conception of a new recreation center with cybernetic climatization and
a green roof (see Fig. 3). In all these cases the cybernetic interpretation clearly showed the difference between critical, active, re-
active or buffering components of the system, while its cybernetics
is revealed by investigating its partial feedback cycles.

Fig. 3. A model of the PUEBLO in Frankfurt,
a new recreation center, conceived
on the basis of a biocybernetic study.

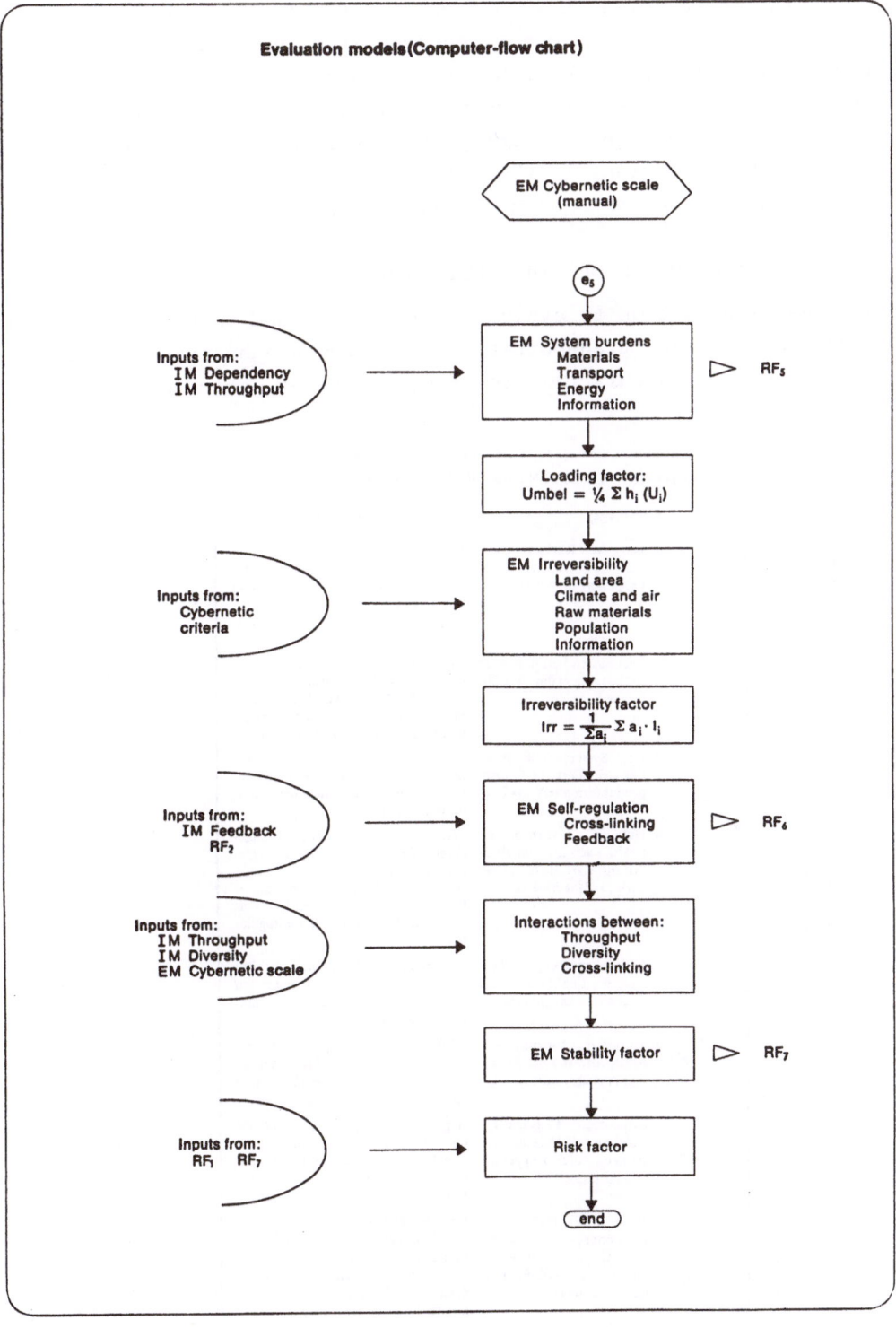

Evaluation models (Computer-flow chart)

EM Cybernetic scale
(manual)

e_5

Inputs from:
IM Dependency
IM Throughput

EM System burdens
Materials
Transport
Energy
Information

$\triangleright$ RF$_5$

Loading factor:
Umbel = ¼ Σ h$_i$ (U$_i$)

Inputs from:
Cybernetic
criteria

EM Irreversibility
Land area
Climate and air
Raw materials
Population
Information

Irreversibility factor
$Irr = \dfrac{1}{\Sigma a_i} \Sigma\, a_i \cdot l_i$

Inputs from:
IM Feedback
RF$_2$

EM Self-regulation
Cross-linking
Feedback

$\triangleright$ RF$_6$

Inputs from:
IM Throughput
IM Diversity
EM Cybernetic scale

Interactions between:
Throughput
Diversity
Cross-linking

EM Stability factor

$\triangleright$ RF$_7$

Inputs from:
RF$_1$ RF$_7$

Risk factor

end

Now to the third level, that of evaluation presented in the flow diagram on page 17. Among other features like systemic risk and stability, evolutionary possibilities and other criteria of survival, there are those of structural and dynamic organization, expressed in a package of overlapping rules, which contains somehow all the other features. In the following I therefore will list and comment these rules in a sort of checklist.

IV.   THE EIGHT RULES OF CYBERNETIC ORGANIZATION

This checklist allows to evaluate our actions and interferences, our production methods, transportation or energy use against the principles of ecosystems. And this forms the functioning of entire spheres of

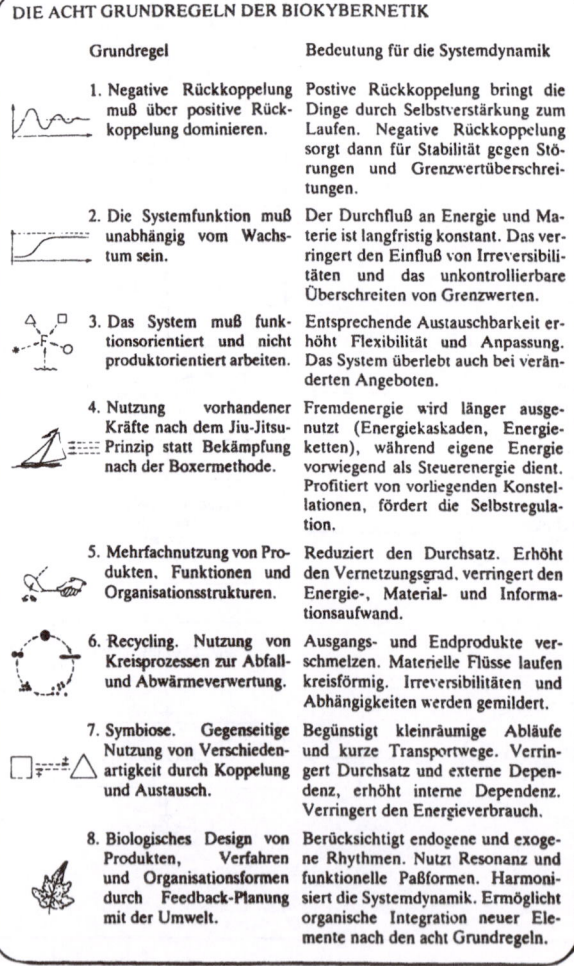

DIE ACHT GRUNDREGELN DER BIOKYBERNETIK

Grundregel	Bedeutung für die Systemdynamik
1. Negative Rückkoppelung muß über positive Rückkoppelung dominieren.	Postive Rückkoppelung bringt die Dinge durch Selbstverstärkung zum Laufen. Negative Rückkoppelung sorgt dann für Stabilität gegen Störungen und Grenzwertüberschreitungen.
2. Die Systemfunktion muß unabhängig vom Wachstum sein.	Der Durchfluß an Energie und Materie ist langfristig konstant. Das verringert den Einfluß von Irreversibilitäten und das unkontrollierbare Überschreiten von Grenzwerten.
3. Das System muß funktionsorientiert und nicht produktorientiert arbeiten.	Entsprechende Austauschbarkeit erhöht Flexibilität und Anpassung. Das System überlebt auch bei veränderten Angeboten.
4. Nutzung vorhandener Kräfte nach dem Jiu-Jitsu-Prinzip statt Bekämpfung nach der Boxermethode.	Fremdenergie wird länger ausgenutzt (Energiekaskaden, Energieketten), während eigene Energie vorwiegend als Steuerenergie dient. Profitiert von vorliegenden Konstellationen, fördert die Selbstregulation.
5. Mehrfachnutzung von Produkten, Funktionen und Organisationsstrukturen.	Reduziert den Durchsatz. Erhöht den Vernetzungsgrad, verringert den Energie-, Material- und Informationsaufwand.
6. Recycling. Nutzung von Kreisprozessen zur Abfall- und Abwärmeverwertung.	Ausgangs- und Endprodukte verschmelzen. Materielle Flüsse laufen kreisförmig. Irreversibilitäten und Abhängigkeiten werden gemildert.
7. Symbiose. Gegenseitige Nutzung von Verschiedenartigkeit durch Koppelung und Austausch.	Begünstigt kleinräumige Abläufe und kurze Transportwege. Verringert Durchsatz und externe Dependenz, erhöht interne Dependenz. Verringert den Energieverbrauch.
8. Biologisches Design von Produkten, Verfahren und Organisationsformen durch Feedback-Planung mit der Umwelt.	Berücksichtigt endogene und exogene Rhythmen. Nutzt Resonanz und funktionelle Paßformen. Harmonisiert die Systemdynamik. Ermöglicht organische Integration neuer Elemente nach den acht Grundregeln.

life, right down to individual firms, consumer behaviour, governmental measures of the design of individual products.

1.   Self-regulation by Dominance of Negative Feedback.

A feedback control system stabilises itself via negative feedback. This mode of self-control in circulatory processes or between limit values is the most important organizational principle of a sub-system if this is to survive within the overall system.

I want to illustrate this with an example of predator and prey. The faster the wolf runs, the more hares he can catch and eat. The more hares he eats, the fatter he gets and its running becomes slower. The slower he runs, the less hares he catches, the thinner he gets, the faster ge can run again, catch more hares, become fatter, and so on: negative feedback.

Positive feedback is necessary as well in order to get things started. It is the motor within the system. However, every sub-system which changes permanently to positive feedback (circulus vitiosus) will enter into a process of amplification in one direction or the other, i.e. it will either explode or freeze. In either case it destroys itself, thus eliminating a disturbing element in the overall system.

Therefore any positive feedback cycle has to be dominated by a negative feedback, bringing it back to an equilibrium by self-regulation. At least that is the cheapest way of governing - even in environmental management.

2.   Independence of Growth.

The settling down of a system to a stable equilibrium is not compatible with continuous growth of this system. In biological processes, one always finds either growth alone (unstable, temporary) or a functioning (stable, permanent).

If a sub-system such as a cell of the brain is completely differentiated and has ceased to grow completely, it functions optimally. Every system, every process, every product therefore should be checked to determine whether it is not primarily growth- but function-orientated.

Not growth as such, but dependence on growth is dangerous. One will deviate from the logistic curve and instead of turning into a new equilibrium the system will collapse (see Fig. 2) as it was the case from 1981 to 1982 for some countries in South America.

3.   Independence of the Product.

The functionally-compatible differentiation of each "cell" of a system simultaneously means that the several products which are formed to meet the needs of permanent functioning are temporary and thus secondary: Products come and go, function is permanent.

The mitochondria for example, minute power stations in the cells of the human body, have the task of controlling the conversion of matter and energy. Using one and the same cycle, they can process carbohydrates to carbon dioxides or change over to the the production of amino acids. A principle that is typical of all biological circulatory processes, from the smallest to the largest.

Taking this viewpoint, the Volkswagenwerk should not understand
itself as automobile construction industry but as being in the
traffic business; electricity companies should not think of themselves as power generators but as energy suppliers, something
that can also imply the obligation to reduce power demands or
to replace energy consumption by alternatives. One of the first
companies that understood this was the P.P.&L.

4. The Jiujitsu-Principle Instead of the Boxing Method.

This is to utilize already-existing forces and energies and to
control and divert these in the desired direction with almost
no own energy. By means of energy cascades, energy chains and
energy coupling which observe this principle, nature achieves
an incomparably high degree of energetic efficiency.

5. The Principle of Multiple Use.

Viable systems show a preference for products and processes with
which they can kill two (or even more) birds with one stone -
in principle a variation on the jiujitsu theme.

6. The Principle of Recycling.

The principle of recycling is strengthening the realization of
the previously stated rules: the beneficial re-integration of
waste products (a term which is completely foreign to nature)
into the living circulatory process of the participating systems.

This calls for a departure from the unlinked, mono-dimensional
line of thinking to which we have been educated, a line of thinking that knows only beginning and end, definite causes and effects.
In a circulatory process, the difference between base material
and waste disappears in the same way, in which cause and effect
merge in a cybernetic feedback system.

7. The Principle of Symbiosis.

Symbiosis is the coexistence of differing species to their mutual
benefit. In biology one finds widely-varying forms of symbiosis
- from the case of ants milking aphids, being protected and fed
by them, over our intestinal bacteria, which live off man's food
giving him vital vitamins in return, to the global "open" symbiosis
between the animal and vegetable world via the circulatory system
of photosynthesis and respiration by chloroplasts living in
symbiosis with the plant cell.

Symbiosis always leads to considerable raw material, energy and
transport savings for all participating elements and thus to
multiple, usually free benefits. The more differences there are,
the more possibilities exist for symbiosis. Symbiosis is therefore favoured by diversity within a small space. Large uniform
structures, central energy supplies, monocultures of industrial
areas, in agriculture or in products design,or pure dormitory
towns must therefore manage without the advantages of symbiotic
relationships, and thus without their stabilizing effect.

Profiting from symbiosis therefore means: small space units when
planning anew, but also a sensible coupling of all existing in-
stallations, for example, in the industrial sector. One can go
far beyond the function of "waste material exchanges" and form
a sort of ecosystem in industry: metal processing enterprises
that cooperate with papermills or breweries, a construction
materials industry connected with coal desulphurization, a food

industry with connected water purification and waste utilizations and new plants selected in order to form the missing links in the chain. The development of symbiosis is, however, primarily a communication task; the technological aspect is secondary.

8. Basic Biological Design.

The final rule to be stated here concerns itself with organizational cybernetics and planning and with creative bionics. Every product, every function and organization should be compatible with the biology of man and nature, which involves already an organic planning by feedback with the environment - for instance with the social environment by participation of citizen groups.

This is not only an ecological requirement, but is steadily becoming an economical requirement also. Environmental problems teach us better management in general, meaning true progress. When rivers lose the ability to purify themselves, this represents just as sudden a financial burden as, for example, when humans lose their immunity as a result of stress, thus leading via sickness and reduced efficiency to heavy social burdens.

V.   CONCLUSION

These rules therefore apply for single cells, for multiple cells, for multiple cell organisms and just as well for populations and ecosystems. All this makes the biospere what it is. An absolutely unique superfactory that controls and regulates itself, that has withstood all external influences, and that has already achieved the sensational age of several thousand million years.

I am convinced we no longer can act as thoughtless as we did. Lester Brown, Chairman of the World Watch Institute in the United States first spoke out the alarming result of recent studies: that our society is reaching the border of a world-wide economic crisis, definitely caused by the destructive exploitation of nature, thus reducing below the minimum its vital resources and services - vital for our species.

For industry and economy this means a redefining of their tasks and a gradual reorganization to fit into existing environmental conditions: in controlling, in building-cybernetics, in management-teaching, in developing-aid and other areas; in the new way of bio-cybernetic thinking worked out by different schools beside of my own: by Joel De Rosney in France[6], by Edward Goldsmith in Britain[7], by Hermann Haken[8] and Dietrich Dörner[1] in Germany or by the economists Hans Ulrich, Fredmund Malik and Gilbert Probst in Switzerland[9]. Thus, biocybernetic thinking has entered the first fields and faculties, and its practical implications have been published by different authors. There are already a few firms who have understood this need and who work along the mentioned eight rules. These firms, I am sure, don' t have to fear as much as others

the economic crises to come, and the moment they don't relay on econo-
metrics but switch to evolutionary management, working no longer against
our biosphere and its eternal rules, but with it. I close with a word
of Francis Bacon: In order to govern Nature, one must obey her.

REFERENCES

1   D. DÖRNER: Problemlösen als Informationsverarbeitung.
     Kohlhammer, Stuttgart 1976.   D. DÖRNER: Wie Menschen eine Welt
     verbessern wollten - Ein psychologisches Experiment. Bild der
     Wissenschaften 12, 48 (Februar 1975).

2   M. CLARKE: The Development of an Environmental Simulation Game.
     Working paper 208, School of Geography, University of Leeds,
     Nov. 1977; F. VESTER: Ökopoly - ein kybernetisches Umweltspiel.
     Zu beziehen durch Studiengruppe für Biologie und Umwelt GmbH,
     Nußbaumstr. 14, 8000 München 2 (1983).

3   G. PROBST und F. MALIK: Evolutionäres Management. Die Unternehmung
     (Schweizerische Zeitschrift für Betriebswirtschaft) 35, 121 (1981).

4   F. VESTER: Urban Systems in Crisis. Deutsche Verlagsanstalt,
     Stuttgart 1976. Pocket-book edition (German only) dtv, München
     1983.

5   F. VESTER and A. v. HESLER: Sensitivity Model.
     Zu beziehen durch Umlandverband Frankfurt, Am Hauptbahnhof 18,
     6000 Frankfurt 1.

6   J. DE ROSNEY: Le Macroscope - vers une vision globale.
     Edit. du Seuil, Paris 1975.

7   E. GOLDSMITH et al.: A Blueprint of Survival. Tom Stacey, London
     1972; see also many articles in "The Ecologist".

8   H. HAKEN: Synergetics. Springer, Berlin 1978; D. DÖRNER: Problem-
     lösen als Informationsverarbeitung. Kohlhammer, Stuttgart 1976.

9.   H. ULRICH: Management - eine unverstandene gesellschaftliche
     Funktion. In: H. SIEGWART u. G. PROBST (Hrsg.): Mitarbeiter-
     führung und gesellschaftlicher Wandel, S. 133 ff., Paul Haupt,
     Bern 1983; F. MALIK: Zwei Arten von Managementtheorien: Kon-
     struktion und Evolution. Ibid. S. 153 ff.; G. PROBST: Kyberneti-
     sche Gesetzeshypothesen als Basis für Gestaltung und Lenkung im
     Management. Paul Haupt, Bern 1981.

PHYSICAL AND TECHNOLOGICAL RESTRICTIONS OF VSLI

I. Eisele

Institute of Physics, Faculty of Electrial Engineering
Federal Armed Forces University Munich, FRG

INTRODUCTION

Talking about parallel processing one usually compares the capability of a computer with the human brain. In general, however, this is not a good comparison because the computer has a completely different architecture and is based on different physical principles. Depending on these properties onw or the other is more suitable for solving a particular problem. Considering a classical equation for neuron nets /1/ one can show the fundamental differences between the two systems "brain" and "computer":

$$U_h(t+\tau) = \theta \sum_k a_{hk} U_k(t) - S_k \qquad (1)$$

where h is the number of a neuron, θ the step function, τ the switching time, k the number of connections, a_{hk} the coupling constants and S_k the threshold. Working with Boolean logic the coupling constants are "0" and "1".

In Table 1 the parameters for the two systems are shown.

	BRAIN	COMPUTER
h	$\approx 10^{11}$ neurons /2/	$\approx 10^8$ bit
k	$< 10^4$	$\leqslant 32$
$\tau\,[s]$	$\approx 10^{-2}$	$\approx 10^{-7}$

Table 1. Estimate of the total number of elements, connections per element, and switching time.

Even comparing these numbers is quite difficult because a neuron can be a very complex unit (corresponding for instance to a microprocessor) which operates on a stochastic basis whereas a bit is just a digital memory on a deterministic basis.

Besides the different logic operation (stochastic and deterministic, respectively) there is another big difference which is not apparent immediately: it is the switching energy for the information transport. For the brain the transport is carried out by ionic and chemical reactions. The corresponding power losses are very small. For electronic circuits the information per bit is carried by some 10^5 electrons. Due to their recombination behaviour a power loss occurs which has to be dissipated within the material. Because the maximum power dissipation is about 1 Watt per chip, this strongly limits the number of devices on a chip as well as the clocking frequencies.

The aim of the computer development in the past was directed towards faster data transport and larger information capacity. The necessary algorithms for the serial configuration have also been developed and proved to work well for many problems. In fact, the computer very often is much faster then the human brain. Only recently it has been seriously tried to solve problems such as image processing with the help of a computer. For such associative processes parallel configurations are much better suited and in this case the human brain is superior to the computer. In the following some aspects are given which show the technological restrictions with respect to a parallel data processing of microprocessor systems.

LIMITATIONS OF INTEGRATED CIRCUITS

As can be seen from Table 1, one has to discuss mainly switching speed, integration density, and the number of connections for very large scale integrated (VLSI) circuits. The circuitry is limited to single chips fabricated in planar technology. This means that device size and position on a chip are basically two-dimensional problems. Furthermore, the high clocking frequency as well as the limited number of connections hint that for the computer it is useful to replace spatial connections by time steps, i.e. use serial instead of parallel data processing.

Maximum Geometrical Device Density

Many papers have been written about this problem /3,4/. For an extensive study one has to consider technological as well as physical restrictions. However, it should be noted that all of these estimates consider planar devices and the third dimension into the bulk of the semiconductor material can be neglected. For the classical MOS (metal-oxide-semiconductor) transistor, as shown in Fig. 1, one can see that

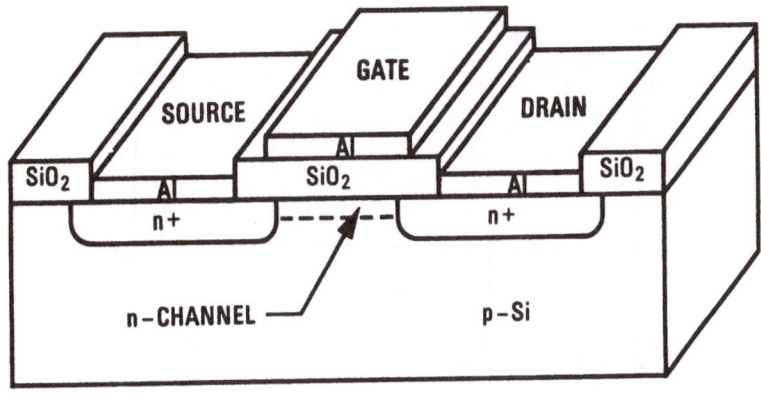

Fig. 1.   n-channel MOS transistor.

the length of the channel determines the active device. In the "Off" state two $p-n^+$ junctions are build up and in order to avoid a cross-talk between them, the minimum distance has to be twice the depletion layer width. It depends on doping concentration and applied voltage as can be seen from Fig. 2. The doping concentration cannot exceed $5 \times 10^{17} cm^{-3}$ because otherwise a hopping conduction due to the impurities occurs. On the other hand the bias potential commonly amounts to 5 V. The resulting depletion width is about 0.2μm and the total channel length >0.4μm. This is a conservative guess but for all estimates values >0.1μm are obtained. The width of the two ohmic contacts (n^+ regions) is mainly determined by technological procedures and is usually

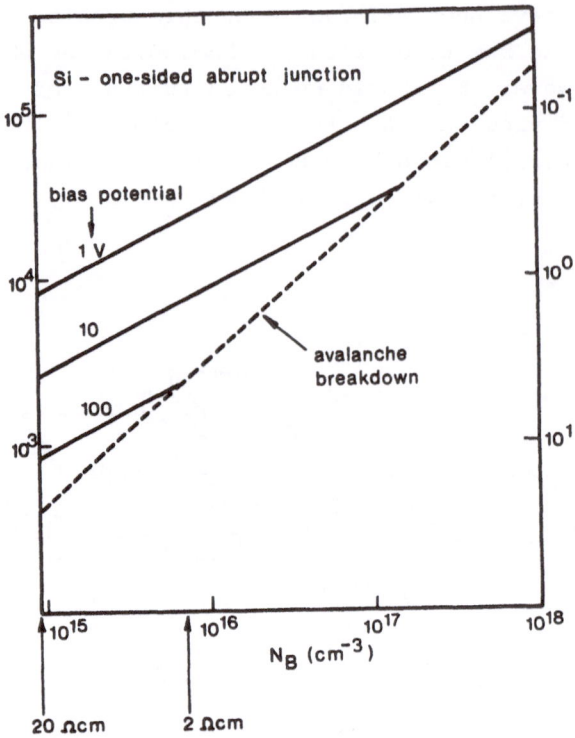

Fig. 2. Depletion layer behaviour as a function
of doping concentration and applied
voltage.

comparable to the channel length. Therefore the total minimum length
of a MOS transistor might be 1.2μm. Considering the same value for the
insulation between neighbouring devices one obtains an area of 1.4x1.4μm^2
and therefore a maximum geometrical integration density of 5x10^7 tran-
sistors/cm^2. However, this is not a realistic value because no space
for connections has been left and they limit our circuits significantly
as will be shown.

Switching Time

As already mentioned the information is represented by an electronic charge which has to be transported through an active device or along a line. The physical limitation of this transport is the speed of the charge carriers. For a semiconductor it is known that this speed saturates with increasing electric field /5/. For silicon this value amounts to $V_{sat} \approx 10^7$ cm/s at a temperature of 300 K. If we consider an effective device length of 0.5μm the drift time of electrons amounts to:

$$t_d = \frac{L}{V_{sat}} \approx 5 \cdot 10^{-12} \; [s] \qquad (2)$$

This is the ultimate value of the switching time which is several orders of magnitude faster then the value in Table 1. However, the time according to equ. (2) is the "internal" device time which, after changing the voltage, is necessary to follow the new conditions. After this process the charging or discharging of the functional or parasitic capacitances of the circuit begins.

It is quite difficult to calculate RC terms because they depend strongly on the specific design of the circuit. Therefore the following values can only be a very crude estimate. For a MOS inverter stage we can assume an effective load resistance which corresponds to the resistance of a transistor and is of the order of 1 kΩ. The minimum capacity for a transistor with an area of 0.5×10μm^2 and an oxide thickness of 20 nm is 8.4×10^{-15}F. Multiplying this value by a factor of two to account for connections one obtains a minimum time:

$$t_{RC} \approx 39 \times 10^{-12} \, s \qquad (3)$$

here the time t_{RC} corresponds to the point where 90% of the capacity are charged.

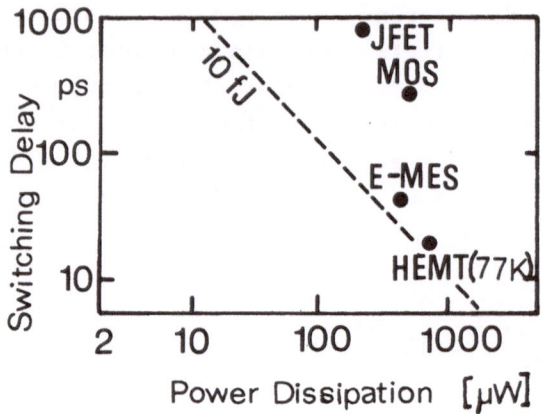

Fig. 3. Switching delay versus power dissipation
for different single devices and technologies.

It should be noted that these are ultimate values which are not
achieved until now. However, from Fig. 3 it can be seen that by de-
creasing the temperature and by using GaAs-GaAlAs hetero-junction
structures (HEMT: high energy mobility transistor) /6/ the switching
time for a single device can be reduced significantly.

At this point it is useful to discuss the charge losses by
scattering and recombination of charge carriers. The resulting loss
energy converts into heat and has to be dissipated.

For instance, the loading of a memory device is determined by the
current transport. A memory device is usually represented by a capacity
which has to be charged or discharged in order to describe a logical
"0" or "1". The totally accumulated charge Q amounts to Q = CV where
C is the technological realized capacitance and V the applied voltage
which for common integrated circuits is 5 V. The charge will be accumulated according to $Q = J dt_s$, where J is the current and dt_s the

switching time. This means that shorter switching times can be achieved by larger currents or smaller load charges, i.e. capacitances. The latter value cannot be reduced significantly because otherwise the signal would be buried within noise.

For thermal equilibrium the above considerations yield the well-known power delay product:

$$Pdt_s = CVdV \qquad\qquad (4)$$

where P is the power which has to be dissipated, dt_s the dynamic switching time, C the capacitance to be charged, V the supply voltage and dV the accelerating voltage for the charge.

For a single device this factor can be taylored appropriately but for very large scale integrated (VLSI) circuits it correlates the number of devices to the switching speed, i.e. the clocking frequency. If no artificial cooling is introduced the maximum power dissipation per chip is approximately 1 Watt. Let us assume that the static power dissipation is negligible, i.e. only the dynamic charging or discharging of a device according to equ.(4) is important. For n devices on a chip and a clocking frequency f_c one then obtains:

$$P <1 \text{ Watt} = f_c nCVdV \qquad\qquad (5)$$

This means, that for large clocking frequencies automatically the number of devices per chip is reduced. For todays technologies the relationship is demonstrated in Fig. 4. It clearly shows that for highly integrated circuits we are far away from the theoretical switching time of a single device which is of the order of 10^{-11} s. So far we only discussed active devices and disregarded connections between them.

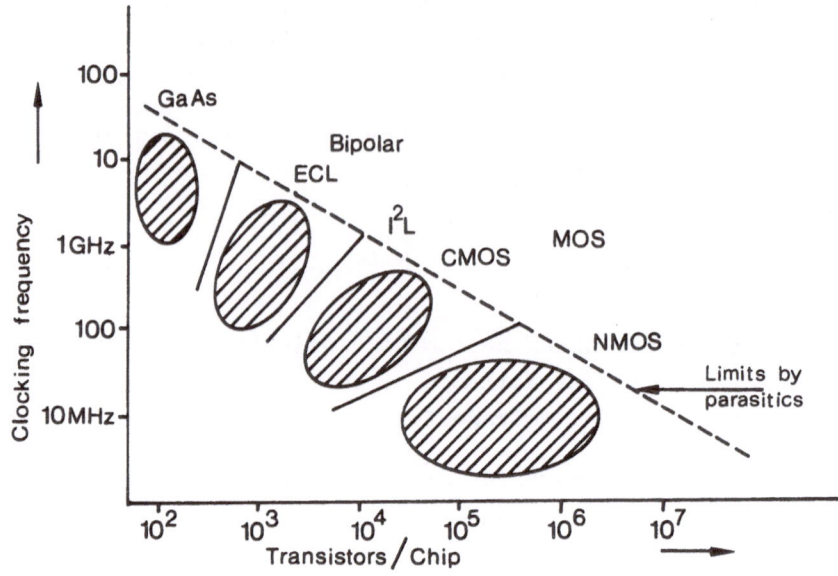

Fig. 4. Clocking frequency versus total
number of devices on a chip.

Connections

The number of connections between different elements on a chip
is basically determined by the available technology. For all VSLI
circuits we have to recall that we use a planar technology, i.e. the
circuitry has to be designed two-dimensionally. The thickness of the
single crystalline material of a chip is about 400µm whereas the
active layer useable for devices amounts to approximately 6µm. The
conductive channel region of a MOS transistor extends even only 10nm
into the semiconductor material. From this point of view it becomes evident
that within the semiconductor material there cannot be too many layers
containing connections between devices. Next, one could argue that
above the semiconductor surface there is a lot of space which could
be used up for connections. However, each line has to be realized
via a photographic process which always contains an insulating SiO_2
layer of the order of 1µm thickness and a technology process with a

temperature of about 1000°C. As a result we obtain quite complicated three-dimensional structures as can be seen from Fig. 5 for the example of a CMOS (complementary MOS) device. As far as the connections

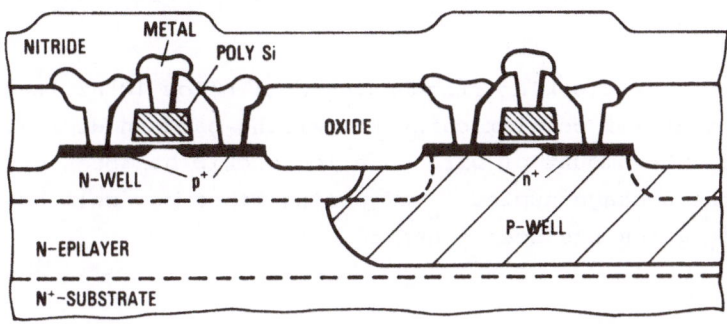

Fig. 5. Structure of a CMOS device in planar technology.

are concerned we can distinguish three different planes: (a) the diffusion region denoted by the highly doped n^+ and p^+ regions, (b) the highly doped polycristalline layer "poly-Si", and (c) the metallization which in general consists of aluminum and has to be the last step because aluminum cannot withstand a technology step of 1000°C. In addition the most advanced technology uses a poly-Si1 and a poly-Si2 process and thus we obtain a total of 4 layers which can be used for connections. This means that not always the shortest distance between two points can be used and the average length of the signal lines increases. This again will limit our circuits as will be shown next. The maximum signal speed in a solid is given by $v_{max} \approx c/\varepsilon_r$, where c is the velocity of light in vacuum and ε_r the relative dielectric constant of the material. For silicon this value will be further reduced by including losses in the material /7/ and one obtains $v_{max} \approx 3 \times 10^9$ cm/s. If we assume that at a certain device all incoming pulses have to arrive within 0.1 of the clocking time we obtain a maximum length for the connecting lines:

$$L_{max} < \frac{0.1 v_{max}}{f_c} \qquad (6)$$

and consequently the maximum chip area is limited to L_{max}^2 . For $f_c = 10^8$ Hz one obtains L_{max}^2 is $9\,cm^2$ whereas for $f_c = 10^{10}$ Hz the value of L_{max}^2 is $9 \times 10^{-4}\,cm^2$. For todays chips where we have areas of about $0.4\,cm^2$ and clocking frequencies of $10^7 - 10^8$ Hz these limits have not been quite reached but by further increasing the switching speed we will run into serious problems.

But even for clocking frequencies well below 10^8 Hz it must be recognized that for complex chips the wiring pattern dominates the design area. The amount of wire on a logic chip has been estimated /8,9/ and the average number N_L of wiring lines per cell on a chip with n logic gates has been found to be:

$$N_L = kn^p \qquad\qquad (7)$$

where k is the average number of terminals per gate and p(0<p<1) is a constant for a given logic graph. For an area A of one logic cell the total line length L per logic cell in the wire dominated case (where only lines are considered) is given by $L = N_L A^{1/2}$. If the width of a single wire amounts to W and m levels of wiring are available one obtains:

$$L = \frac{WN_L^2}{m} = \frac{W}{m} k^2 n^{2p} \qquad\qquad (8)$$

Usually k is about 6 whereas p strongly depends on the organization of the circuitry. Highly parallel circuits may have exponents p as large as 0.75 /10/ while highly serialized circuitry may be designed with exponents as low as 0.17 /11/. This means that for a chip with a line width of $4\,\mu m$, four layers for connections, and 10^3 logic gates the total length of lines amounts to $3.6 \cdot 10^5\,\mu m$ for 2 p = 1/3. On the other hand for 2p=1 one obtains $360 \cdot 10^5\,\mu m$. The latter value which is close to parallel processing yields a wiring area of $144\,mm^2$. With present technologies this is not achievable.

At this point it is useful to discuss connections from a chip to the "outside world". This is achieved by ultrasonic bonding of thin wires to aluminum pads of $100 \times 100\,\mu m^2$. The pads are placed at the circumference of the chip and thus for a chip size of $5 \times 5\,mm^2$ we obtain a maximum of about 150 connections (considering that there has

to be a certain distance between the pads). Because several lines have
to be used for clocking frequency, supply voltages and so on, the
number of signal connections is further reduced.

If one considers that due to the failure densities of todays
technology the chip size is limited to about $100mm^2$ one can formulate
the following design restrictions:

(a)   The clocking frequency (speed of signal transport) cannot
       be increased arbitrarily because increasing speed limits
       the number of devices per unit area.

(b)   Due to planar technology the number of connections cannot
       be increased arbitrarily because increasing connection
       density also decreases the number of devices per unit area.

(c)   The number of connections is reduced drastically if several
       chips are connected with each other.

VLSI for Parallel Processing

Talking about parallel processing we cannot distinguish between
logic and memory as it is done for conventional computers. It is
necessary to combine many small "microcomputers" or "neurons" to a
net. The difference between serial and parallel processing is demonstrated in Fig. 6. Actually a real device will always be between the

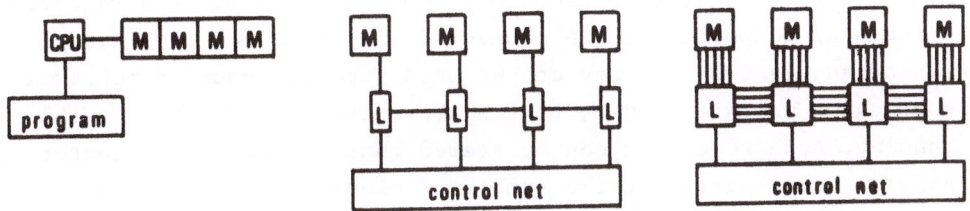

Fig. 6.   Organization of serial and parallel
          processing machines.

two extrema. Because the existing computers in most cases do a very
good job, one has to ask the question in what cases parallel processing is necessary. One growing area is the field of pattern recognition
where for instance in the case of image processing many informations
have to be compared simultaneously. On the other hand parallel processing is necessary if the number of data is too big to be handled
within a reasonable time interval. Examples are predictions in
meteorology and economy models.

Several ways to introduce parallel processing are described in
this book and they all are based on combining relatively large subsystems via nets or bus systems. In this paper it will be discussed
whether it is possible to carry out on chip parallel processing or
at least on wafer processing.

Because the success of parallel processing strongly depends on
the connecting network which in turn is different for each problem,
a general computer seems to be impossible, and therefore for each
problem a different device is necessary. For one-dimensional problems
(for instance the one-dimensional diffusion equation) the connecting network is
relatively simple. Even connecting chips on a wafer is not a serious
problem /12/. However, as soon as two or many-dimensional problems
have to be solved, the connecting network becomes very complex.

Assuming a clocking frequency of 10 MHz and a chip size of 5x8mm^2
we can use the maximum integration density which amounts to approximately
10^7 devices. If a basic unit which consists of memory and logic (neuron)
has 500 elements /12/ this means that 2×10^4 "neurons" can be placed on
one chip. This is tremendous if there would be enough additional planes
for connections. However, as has been shown before, this number is
drastically reduced if more then four connecting layers are required.
Thus, several chips have to be connected and for a two-dimensional
matrix arrangement of the neurons we need several thousand connections
between chips which according to the above estimate is not possible
by bonding. As a result it can be stated that the hindering factor for
parallel processing is not the number of elements but rather the
number of connections. The ideal situation would be a three-dimensional
arrangement as is shown in Fig. 7 where the active elements or neurons
are on top whereas many underlying planes are reserved for connections.

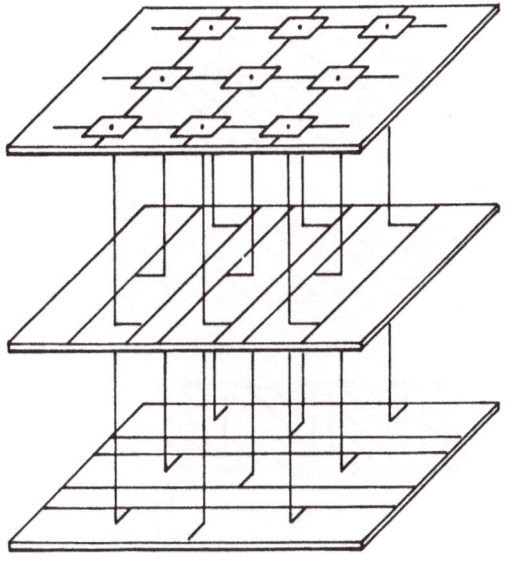

Fig. 7. Three-dimensional connection network.

New Technologies

Recently new technologies have been developed which all have the aim to extend planar technology into the third dimension.

A very promising approach has been undertaken by aluminum thermomigration through silicon (see Fig. 8) and thus connecting different wafers /13/. Aluminum dots are evaporated on the silicon wafer which then is heated to approximately 1000^0C (see Fig. 7). Simultaneously a thermal gradient of 150^0C/cm is applied in a way that the surface containing the droplets is cooler. According to the phase diagram of the Si-Al system a concentration gradient of silicon dissolved in the aluminum occurs. This in turn causes a diffusion of silicon through the aluminum dots. The silicon recrystallizes at the end of the dots.

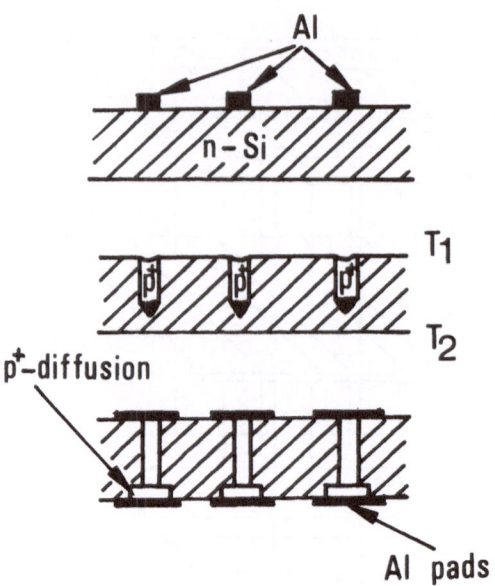

Fig. 8. Schematic diagram of aluminum thermo-
         migration.

After the aluminum has moved through the silicon an Al doped (p-doped)
trace remains. Because the doping concentration is very high ($>10^{19} cm^{-3}$)
one obtains connections with ohmic behaviour. If the material of the
wafer is n-type these connections are isolated against each other through
p-n junctions. The area of the pads which are connected to these
contacts can also be made 100 x 100μm^2 and thus the number of connections
to the outside world can be increased significantly. In fact a micro-
bridge system has been developed /14/ which allows to stack many
wafers on top of each other. The resulting three-dimensional devices
have been built and their performance with respect to image processing
has been proved to be superior to serial computers /14/.

Another new technology which might allow to extend circuits into the third dimension is "Molecular Beam Epitaxy", i.e. the growth of expitaxial layers on a substrate in ultra high vacuum /15/. In this case arbitrary doping profiles can be achieved perpendicular to the surface /16/. The creation of many p-n junctions opens up the possibility to build three-dimensional devices which in turn allow more space for connections. However, there are two problems which are not solved so far: (a) selective epitaxy has to be used in order to restrict the active regions to local areas, and (b) it is very difficult to fabricate connections to the different layers. As an example for a three-dimensional device the following paper will discuss the possibility for the fabrication of a three-dimensional shift register. Despite the fact that such devices seem to be realistic, a lot of technological problems have to be solved and questions such as the power dissipation have to be answered.

CONCLUSION

After estimating clocking frequencies and the number of possible connections for VSLI chips it turns out that with the common quasi two-dimensional planar technology it is not possible to build complete nets large enough for parallel processing. As a result, only bus systems with several parallel working subsystems can be fabricated today. However, by extending the technology into the third dimension, new connection networks or three-dimensional device structures might be possible which allow rethinking the verification of large nets for parallel processing.

REFERENCES

/1/   E.R. Caianiello, Outline of a Theory of Though Processes
        and Thinking Machines, J. Theor. Biol. 1, 204 (1961).

/2/   R. Sint, Neurobiologie und Gedächtnis, p. 26, Fischer Verlag
        (1979).

/3/   B. Hoeneisen and C.A. Mead, Limitations in Microelectronics:
        Bipolar Technology, Solid State Electron. 15, 891 (1972).

/4/   B. Hoeneisen and C.A. Mead, Fundamental Limitations in Micro-
        electronics: MOS Technology, Solid State Electron. 15, 819
        (1972).

/5/   W. Müller and I. Eisele, Velocity Saturation in Short Channel
        Fieldeffecttransistors, Solid State Comm. 34, 447 (1980).

/6/   M. Abe, T. Mimura, N. Yokoyama, K. Suyama, Advanced Device
        Technology for High Speed GaAs VLSI, Solid State Devices 1982,
        ESSDERC Meeting, Munich, 13-16 Sept., p. 25 (1982).

/7/   O.G. Folberth, Signalfortpflanzung in integrierten Schaltungen,
        Int. Elektr. Rundschau 28, 9 and 28, 29 (1974).

/8/   W.R. Heller, W.F. Mikhail, and W.E. Donath, Proc. 14th Design
        Automation Conf., New Orleans, 20-22 June (1977).

/9/   W.E. Donath, Placement and Average Interconnection Lengths of
        Computer Logic, IEEE Trans. on Circuits and Systems 26, 272
        (1979).

/10/   R.L. Russo, On the Tradeoff between Logic Performance and
        Circuit to Pin Ratio for LSi, IEEE Trans. Comput. 21, 147
        (1972).

/11/   R.W. Keyes, GaAs High-Speed Logic, Int. Symp. on Digital
        Technology, Status and Trends, Oldenburg (München) p. 253
        (1981).

/12/   J.D. Becker and I. Eisele, Computing with Neural Nets: Design
        and Technology, Proc. Int. Workshop on Cybernetic Systems,
        Salerno, 9-12 Dec. (1981).

/13/   R.D. Etchells, J. Griuberg, G.R. Nudd, Development of a Three-
        Dimensional Circuit Integration Technology and Computer
        Architecture, Soc. of Photographic and Instrumentation Engineers,
        282, 64, Washington, April (1981).

/14/   G.R. Nudd, in Image Processing from Computation to Integration,
        Ed. S. Levialdi, Academic Press, in press.

/15/   Y. Ota, Silicon Molecular Beam Epitaxy, Thin Solid Films 106,
        No. 1/2, 3 (1983).

/16/   A. Beck, H. Jungen, B. Bullemer, and I. Eisele, A New Effusion
        Cell Arrangement for Fast and Accurate Control of Material
        Evaporation under Vacuum Conditions, J. Vac. Sci. and Technol. (1984).

DESIGN STRATEGIES FOR VLSI LOGIC

Egon Hörbst, Karlheinrich Horninger & Gerd Sandweg
Corporate Laboratories for Information Technology
Siemens AG
8000 Munich 83
West Germany

ABSTRACT

Some typical problems of VLSI circuits and their solutions with the
help of architectural concepts, circuit design and process technology
are presented. These principles are demonstrated on two experimental
chips fabricated in a research process line. Regular structures for
the control part of a VLSI processor are described in more detail.

TYPICAL PROBLEMS OF VLSI CIRCUITS

A VLSI circuit comprises more than 100 000 transistors (including pla-
ces for transistors in ROMs and PLAs). The density and low power consumption needed for such a large number of transistors is only achie-
vabel with MOS technology. As shall be shown later, the characteristics
of the MOS technology influence the design style. There are a number
of VLSI problems, some of which can be solved by circuit design and suitable architectural concepts.

The most evident problem is managing the complexity. This is the reason why first VLSI circuits were memories. From a point of logic complexity these chips are very simple. You "only" have to solve circuit
design and processing problems. For logic oriented circuits, on the
other hand, the problem is how to reduce complexity. One method is to
use regular modules like RAMs, ROMs, PLAs or slice structures as much
as possible. Additionally it is essential to employ CAD tools extensively.

As a result of the high packing density, one can run into power consumption problems. A way to reduce this problem is to use dynamic techniques e.g. precharged busses. But dynamic techniques can be critical
in timing in some areas and can lead to trouble, especially during testing. For VLSI circuits it is therefore better to use static techniques

where possible and dynamic techniques only in large blocks (e.g. PLAs). The power problem can also be solved by using a low power technology like CMOS.

A severe problem is the limitation of pins. We are able to put a whole system with tremendous computing power on a chip but we have difficulties to get the inputs to and the results from the chip. The transfer to the outside world of the chip is approximately one magnitude slower than the transfer inside the chip. The solution is to broaden up and separate the communication paths but this again is limited by the cost and the mechanical problems of packages with high pin count. Architectural solutions might be structures like systolic arrays or concepts like pipelining and distributed processing.

Another consequence of the narrow communication channel is that the controllability and observability of circuit blocks decreases with increasing integration. This leads to testing problems unless special design techniques for testability or selftest are used.

VLSI chips generally need much area. On the other hand, the yield decreases very significantly with increasing area. In memories this problem is solved by adding redundant elements (spare rows and columns). For logic circuits redundancy and error correction on the chip are still areas of research. During the design phase of large chips it is necessary to use area saving structures and to develop good floorplans.

The next problem is the wiring. In logic oriented circuits most area is not consumed by the active elements but by the wiring between them. Since even the most modern MOS technologies only have two metal layers for wiring (because of their high resistance, polysilicon and diffusion can be used for short connections only) the designer must carefully layout the different subblocks for easy interconnection. In the typical case of VLSI, where logic and geometric structures have to be designed together, there is an additional reason for doing this. Long wires result in large capacitances, large transistors to drive them and therefore long signal delays. The designer therefore has to arrange function blocks that fit together and need only few and short interconnections.

The last typical VLSI problem to be mentioned here is the small production volume. Only memories and some successful microprocessors achieve production quantities of more than 100 000 a year. But the future VLSI market will belong to the coustom ICs produced in small quantities. One solution to this problem is to make low volume production cheap, the other is to reduce development time and thus cost by extensive use of

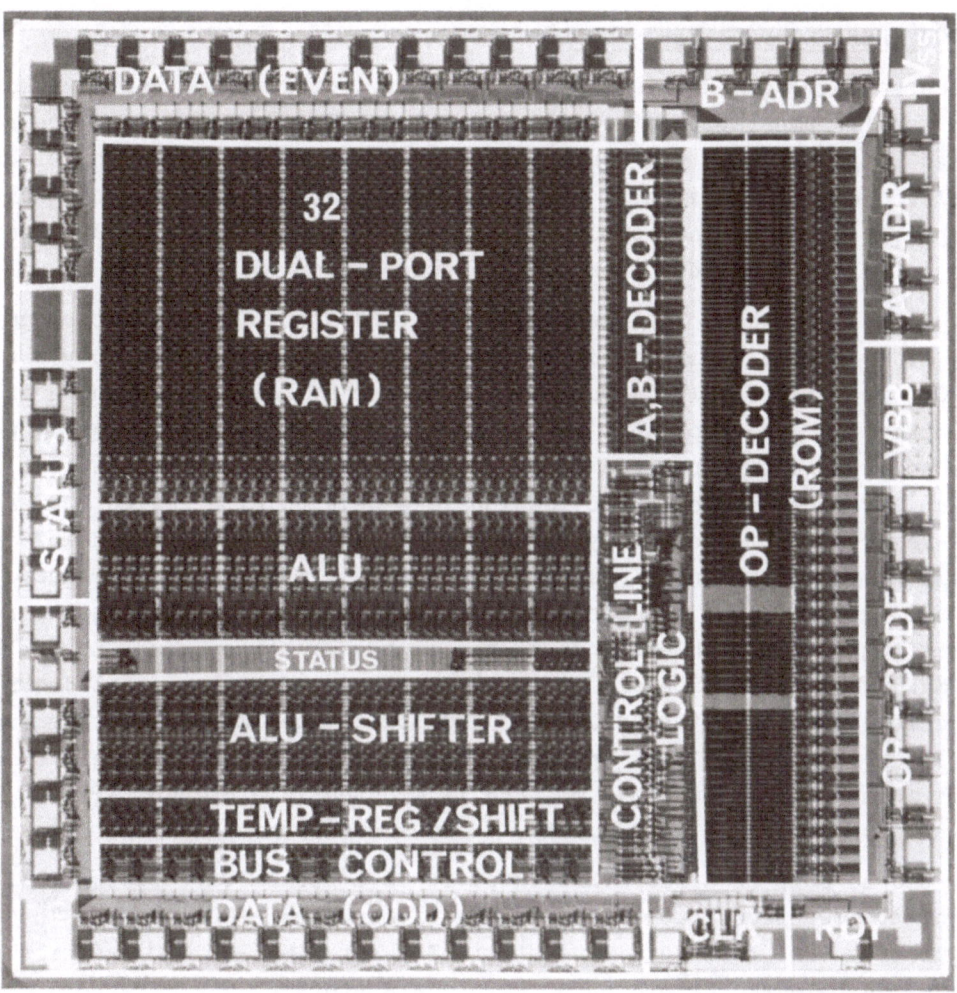

Fig. 1:   Micrograph of the realized 32-bit execution unit

computer aided design combined with architectural concepts suitable
for automation.

STRUCTURES FOR VLSI PROCESSORS

Processor structures have proved to be very powerful and well suited
for the implementation of complex functions. Therefore most of the realized logic oriented VLSI chips have processor structures. Typical examples are microprocessors, peripheral controllers, signal processors,
graphic processors and communication chips. The two chips we have gained our VLSI experience from are processors too.

The first chip is a 32-bit execution unit /1/. When counting the number of transistors, it is not really a VLSI chip because it only has
25 000 transistors. It is rather a model for a VLSI chip. We have tried
to make this execution unit as regular as possible without loosing performance or wasting silicon area. Beside this it was used as a test
circuit for our research fabrication line developing a scaled NMOS sin-
gle-layer poly-Si technology with 2-µm minimum gate length and low-
ohmic polycide for gates and interconnections. The chip was produced
in 1981 and the test results were very satisfactory. The circuit performs logic and arithmetic operations on two 32-bit operands in 125 ns
(8 MHz). Multiplication and division is supported by a special control
circuit to speed up the shift and add logic. Thus multiplication and
division on signed 32-bit operands need only 34 cycles. This execution
unit chip has an area of 16 mm^2 and is mounted in a 64 pin DIP. Its
power dissipation is 750 mW. Fig. 1 shows a micrograph of the chip.

Our second chip is a real VLSI chip. It is a peripheral processor with
about 300 000 transistors and an area of more than 100 mm^2 /4/. The
biggest part of this chip is a static 36-kbit RAM (200 000 transistors)
which can be used for microprogram or data storage. The data word size
is 16 or 8 bit. The instruction format is 32 bits wide. The instruction
set is tuned to the special task of this processor and therefore rather
irregular. Most of the instructions are register-to-register operations
but there are also three-address-operations and operations between external operands.

To realize this chip we have chosen the same technology as mentioned
above except for an additional second metal layer. The design of this
chip is completed and we have got first silicon (Fig. 2).

With both chips we were able to show that for VLSI processors it is a
good architectural concept to partition the processor into an operative

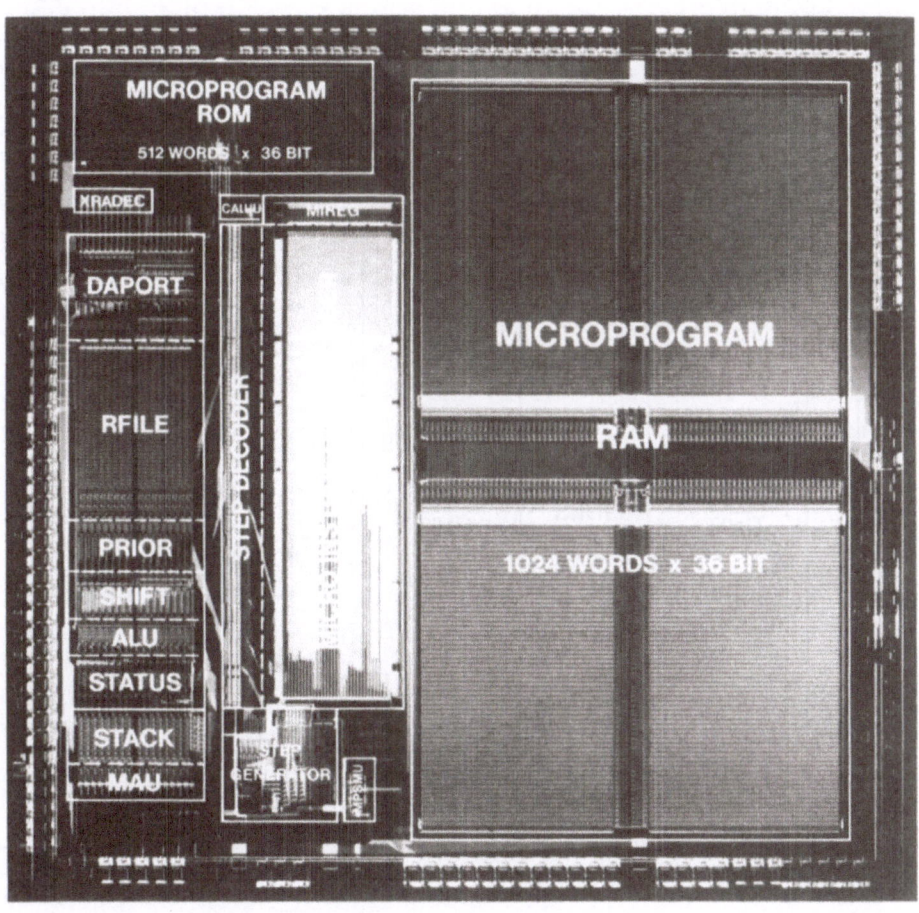

Fig. 2: Micrograph of the realized peripheral processor

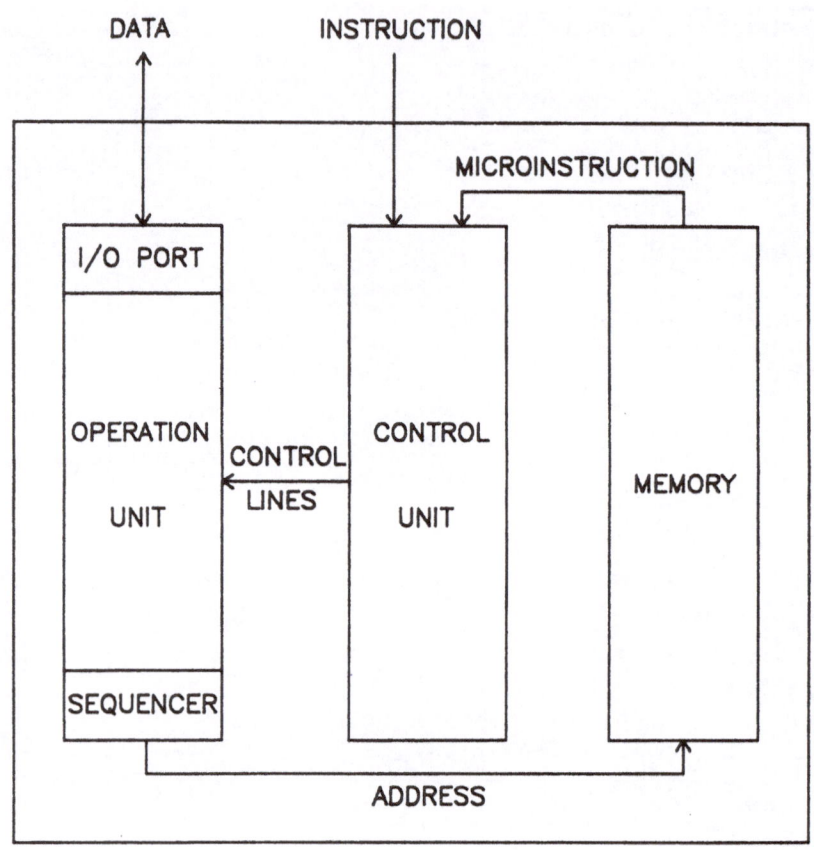

Fig. 3:   Simplified structure of a VLSI processor

part, a control unit, a memory and an I/O unit (Fig. 3). These are the
classical components of a computer. For each of these parts there is an
appropriate design style.

OPERATIVE PART

For the operative part of the processors a slice architecture has proved to be very effective when taking processing speed, area consumption
and design effort into account. The recommended architecture is called
a data path /2/ and consists of several function slices that are built
together without any additional wiring. This is because the cells of the
function slices have an integrated bus system, in our case a 2-bus system (Fig. 4). The data lines run in aluminum to minimize signal delays.
Perpendicular to the data lines, the control lines run in polycide which
has a ten times lower sheet resistance than normal polysilicon. The advantage of running the control lines in polycide is that there is no
change of layers necessary to control the gates of the transistors.

Each function slice is built with bit cells arranged in a serial manner.
So for each function slice only one bit has to be designed. The final
construction of a special data path can then be made with CAD tools.

The slice concept requires a uniform processing width inside the data
path. Various data types with different widths are therefore unfavourable. The width of the data path must be equal to the width of the largest operand handled in one cycle. When shorter operands are loaded into the data path, the additional bits should be filled with the sign
bit or with zeros.

For the processing speed the bus system used in the data path is of vital importance. We have found that a 2-bus system is a very good choice.
A 1-bus system with an accumulator would need one additional cycle for
the very frequent 2-operand instructions. On the other hand the second
bus requires relatively few additional area because the basic cell width
already accomodates 2 power supply lines and at least 4 channels for interconnections. For some special processors the introduction of a third
bus might be advantageous, but in our case there has been no need for
it.

Concerning the electrical operation of the bus system there is the choice between a precharged bus system and a fully static bus system. The
precharge technique has a lower power consumption and may have a higher
speed if it is possible to precharge the busses during phases when no
data transfer occurs. We have chosen this technique in the execution
unit chip because it uses a fixed 4-phase cycle with ample time to pre-

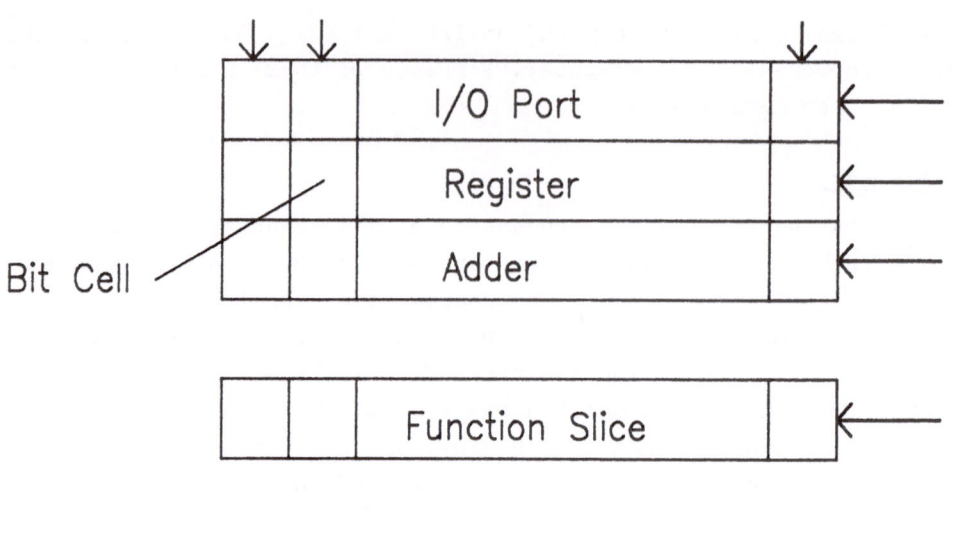

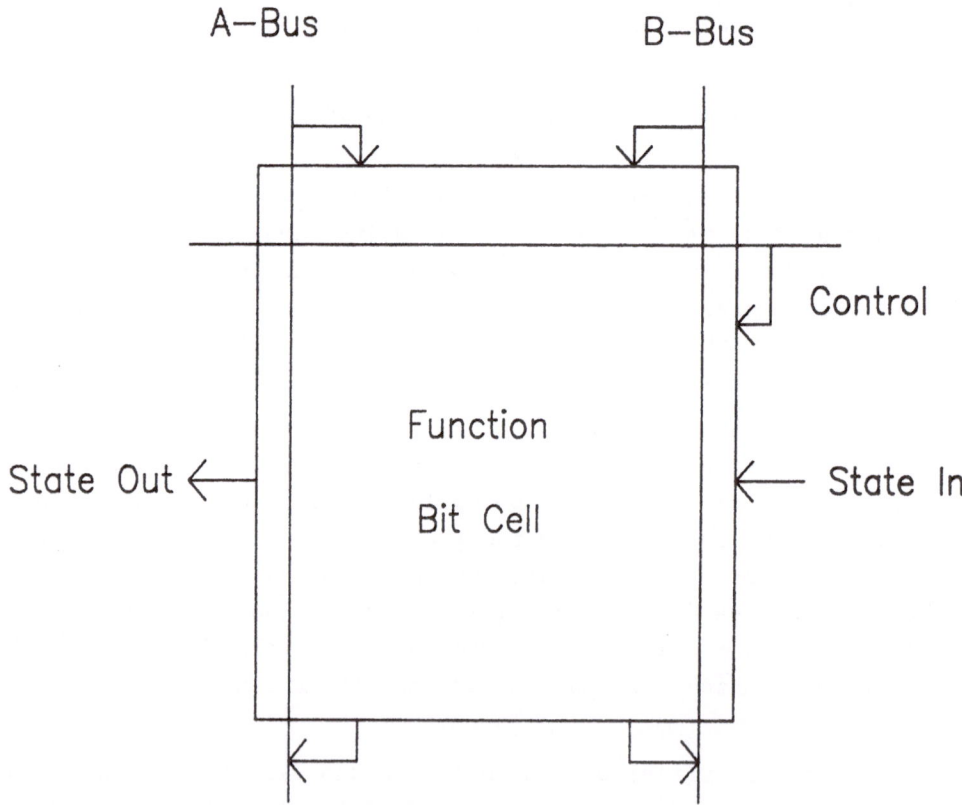

Fig. 4:   Basic principles of the slice technique

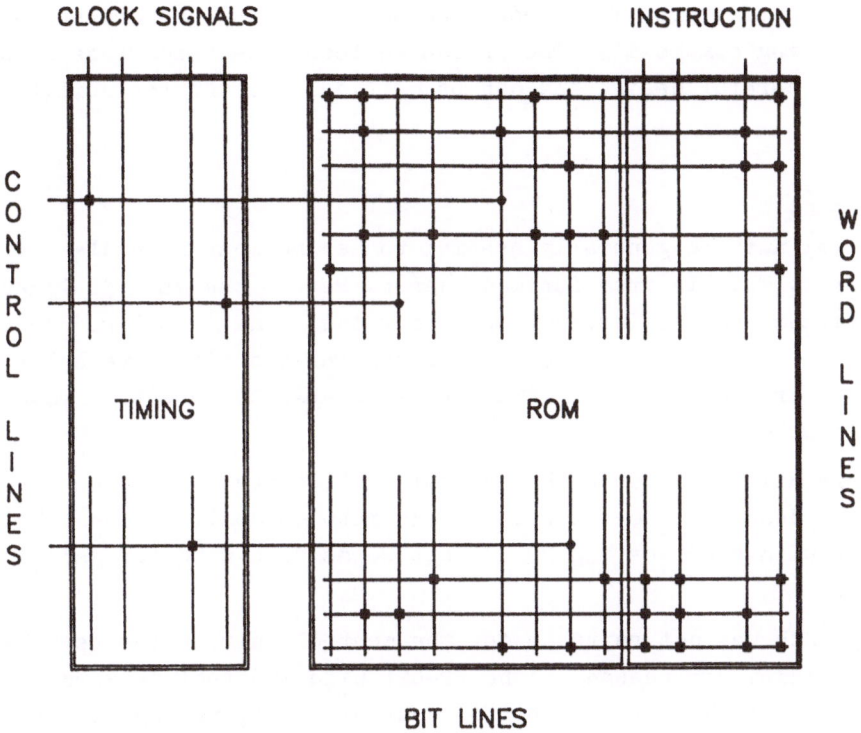

Fig. 5:   Application of a ROM as an instruction decoder

charge the busses. In the peripheral processor chip a variable instruction cycle is used and the timing is therefore more critical. So a fully static bus system has been chosen and each function slice has strong
tri-state bus drivers. This technique needs a little more area but also
reduces timing complexity. The precharge technique needs less area but
also needs careful design and can be critical for the testability of the
circuit.

CONTROL UNIT

The control unit is generally assumed to be the most irregular part of
a processor. This is true for many cases. But if the VLSI designer has
some influence on the instruction format and if he is allowed to spend
a little bit more silicon area, quite regular and effective solutions
for the control unit can be found. We were able to show this with our
experimental chips.

In the execution unit chip the control unit is pretty simple. Its only
task is to decode the 8-bit wide opcode into 40 control lines. This has
been done with a ROM having 208 40-bit words, one word for each opcode
(Fig. 5).

This approach was not suitable for the control unit in the peripheral
processor chip. One reason is the 32-bit wide instruction word, the
other reason is the large number of control lines, namely more than 200.
The obvious solution is to use several small ROMs or PLAs. But we have
found that this would be a large waste of area especially when considering the wiring. We therefore chose a more regular structure which we call
a "degenerated" PLA. Instead of an OR-plane this PLA has only a splitted
OR-line. This is possible if the OR-plane is occupied very weakly as a
result of independent function groups in the instruction format. The AND-
plane then becomes rather long (in our case about 500 product terms) but
its size fits very well to the size of the data path (Figs. 2 & 6).

The output signals of a decoder ROM or PLA can generally not be used for
control lines. They have to be combined with clock signals to produce
the exact timing. This timing stage can be regarded as a second decoder
since it decodes the different phases of an instruction cycle. This second stage adds some flexibility to the decoding scheme especially if a
combination of timing and functional signals is used to clock the control lines. An important advantage of the described decoder structures
is that they can both be generated automatically.

Another task of the control part is to calculate the next instruction

CLOCK SIGNALS INSTRUCTION

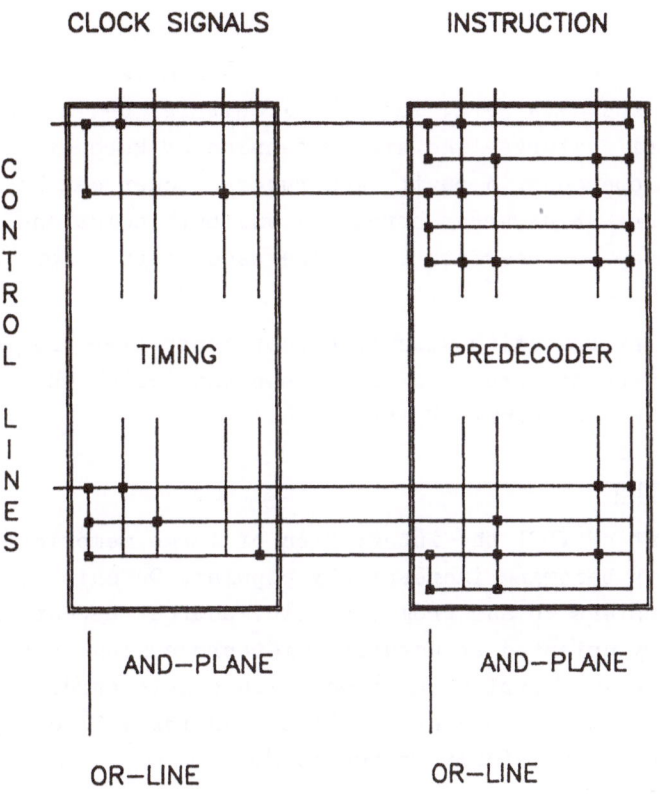

Fig. 6: Application of a PLA as an instruction decoder

address and fetch the next instruction word. This task can be done by a sequencer which can be made to fit very nicely into the slice concept of a data path.

I/O UNIT

The I/O units in our experimental chips are simple and were easily integrated into the data path. The I/O circuit is just a register with some additional features like sign extension or zero extension for the smaller input operands. A parity generator/checker can be realized in a serial structure or an array structure and both solutions fit well into the slice concept of the data path. The same is true for the priority encoder used.

For more complex I/O units with interrupt controller and serial data links it might be necessary to form a seperate block which may even have a processor-like structure of its own.

ON-CHIP MEMORIES

With the advent of VLSI the integration of large memories in logic oriented circuits is becoming increasingly popular. On-chip memories can help to reduce transfers to and from the chip. Another aspect is to replace logic by memory oriented structures. Microprogamming is an established method to implement complex functions with simple hardware. But there are also other memory structures well suited for VLSI designs like e.g. associative memories, stacks or queues /3/.

In additional studies we were able to show that for a small on-chip cache a fully-associative memory structure is very suitable. For a given hit ratio this solution needs the smallest area and has the most regular structure. In the peripheral processor chip we have added an address stack for subroutine calls to the sequencer. This stack is very similar to a dual-port register file but instead of a conventional decoder a pointer in a shift register is used to address the memory cells.

INFLUENCE OF PROCESS TECHNOLOGY

When developing a VLSI circuit the basic characteristics of the technology influences the design styles of the functional blocks. The control PLA of the described peripheral processor is a good example for this. The word lines in a PLA are generally made in polysilicon for a minimum size cell. In the case of the control PLA (Fig. 6) these word lines running vertically are very long and result in a prohibitively high RC time delay (more than one cycle even with the use of polycide). It was therefore essential to run additional lines in the second aluminum layer above the polycide lines and to make contacts to them every 128 product terms. The RC time delay in the PLA word lines was thus reduced to a negligible value. Without this technology feature it would have been necessary to partition the controller.

For the on-chip memory (RAM and ROM) the use of polycide word lines was sufficient to achieve the access times necessary. If only polysilicon would have been available, the memories would have had to be partitioned into smaller subblocks to achieve the same access times and would have resulted in a larger area.

CONCLUSION

Our experience in the development of VLSI circuits has shown that it was necessary to organize the various disciplines like system architecture, circuit design, process technology and CAD in one single team to have close interactions between these disciplines when designing such highly complex chips. As demonstrated in both of our experimental chips, the results of these cooperations were design methods which are very algorithmic and well suited for CAD – and this without wasting area, power or performance.

This work was supported by the Technological Program of the Federal Department of Research and Technology of the Federal Republic of Germany. The authors alone are responsible for the contents.

REFERENCES

/1/ Pomper, M.; Beifuss, W.; Horninger, K.; Kaschte, W.: A 32-bit Execution Unit in an advanced NMOS Technology. IEEE Journal of Solid-State Circuits, Vol. SC-17, No. 3, June 1982.

/2/ Mead, C.; Conway, L.: Introduction to VLSI Systems. Reading, MA, Addison-Wesley, 1980.

/3/ Hoerbst, E.: Case Studies on the Interaction between Process Technology, Architecture and Design Methodology. Conference on Microelectronics, May 1982, Adelaide. The Institution of Engineers, Australia, National Conference Publication No. 82/4.

/4/ Pomper, M; Augspurger, U.; Müller, B.; Stockinger, J.; Schwabe, U.: A 300 K transistor NMOS peripheral processor.
ESSCIRC 1983, Lausanne. Digest of Technical Papers.

CHARGE STORAGE AND CHARGE TRANSFER

IN DYNAMIC MEMORIES

J.D. BECKER

Federal Armed Forces University Munich

D-8014 Neubiberg

I.   INTRODUCTION

Among the many problems parallel processing is faced with - bigger
nets, long range connections, communications with the outside world,
organization, imperfections, etc. - the need for local memories
enjoys a central rôle. To process information means first of all
to get and keep the information where it is needed.

Digitally coded information may be represented in three different
ways: as the contents of memory cells, as logic gates, and as
connections.

Connections seem to be extensively used in the brain; in planar VLSI
technology, however, the number of connections is very limited because
it is only possible to have a very small number of wiring layers (say,
three or four). Even if with the advent of three-dimensional technologies the situation will improve it is not likely to change dramatic-
ally.

Logic gates are fine for a small number of inputs and outputs. For a
larger number of inputs and outputs, however, it is advantageous to
use look-up tables, i.e. memories, instead of logic gates. As an
example, let us mention the use of residue arithmetics in parallel
image processing /1/.

Hence the major part of the necessary information has to be stored
in memory cells.

From a technological point of view we may distinguish between permanent, static, and dynamic cells. In permanent cells, information is represented in a material way (e.g., presence or absence of a link, as in ROMs). In static cells, information is represented as a voltage (e.g., the state of a flip-flop). In dynamic cells, information is represented as a charge package.

Dynamic cells have the disadvantage that they need periodic refreshing, but they have the important advantage that they need only little space, in particular, when being arranged as a shift register.

The shift register replaces most of the connections (which would otherwise be needed) by shifting the data around, thereby saving a lot of space but employing more time. (This strategy may also be found, in a more sophisticated way, in a number of parallel machines, like the Cube Connected Cycles, the Ω Machine, etc.; see /2/ and references quoted therein.)

In this contribution we shall study the physical principles of two types of dynamic shift registers: the CCD (Charge Coupled Device), and the Superlattice Shift Register. Whereas the former is widely being used, the latter could be a very effective storage device in three-dimensional VLSI structures to come.

II. THE CCD SHIFT REGISTER

A static memory cell (flip-flop) requires four transistors, two resistors, and a lot of wiring; hence, it needs a lof of space just to store one bit of information. Consequently, dynamic memory cells have been developed in which information is represented by a charge package. These charge packages are stored in little capacitances.

Fig. 1. Capacitance (left: bipolar, right: MOS)

(For brevity, we shall concentrate on MOS technology in the following examples.) It is still possible to make random access memories this way. A prominent example is the one-transistor cell.

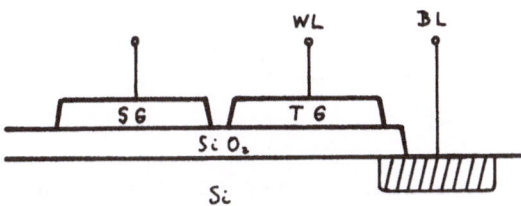

Fig. 2. One-Transistor-RAM Cell

   SG: storage gate, TG: transfer gate

   WL: word line, BL: bit line

For many purposes, however, random access is not strictly required, and we may use CCD shift registers /3/, thereby saving still more space. (One may achieve quasi-addressability by using many small shift registers.) As an example we quote the buried channel CCD shift register.

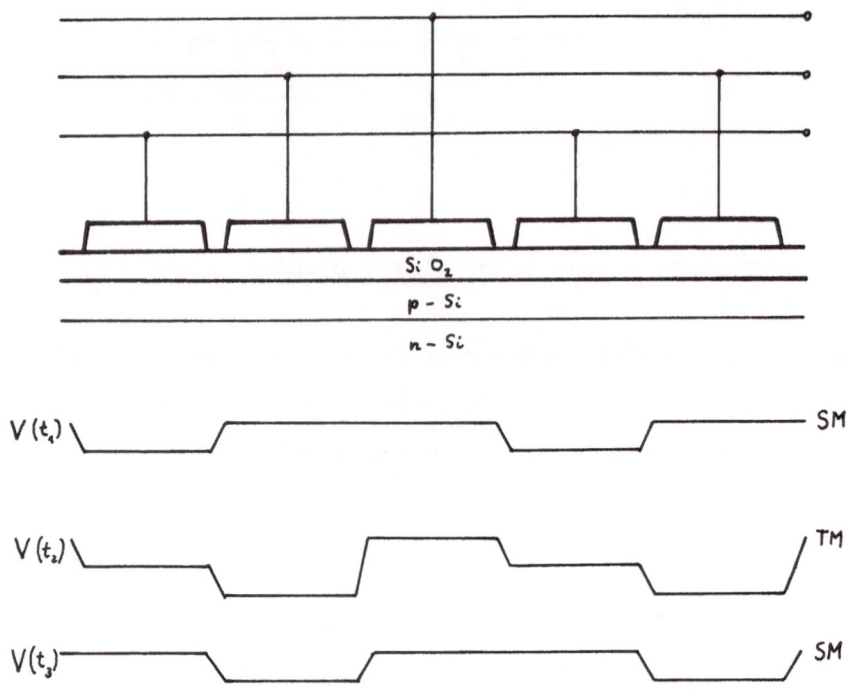

Fig. 3. Buried Channel CCD Shift Register

        Top: structure (schematically)

        Bottom: modes of operation

        SM: storage mode, TM: transfer mode,

        t: time

For the charge transfer, three mechanisms are important: self-induced drift, diffusion, and externally induced drift. Consequently, we may write for the current

$$j = j_{drift} + j_{diffusion} = Q\mu E - D\frac{\partial Q}{\partial x} \; .$$

where $D = \mu kT/q$.

The field consists of two parts:

$$E = E_{ext} + E_{self-induced} = E_{ext} - \frac{1}{C} \frac{\partial Q}{\partial x} \ .$$

The charge transfer is then described by the continuity equation

$$\frac{\partial Q}{\partial t} = - \frac{\partial j}{\partial x} \ .$$

We may calculate the residual charge $Q_r(t)$ of a cell of length L being emptied in a transfer step:

$$Q_r(t) = \frac{1}{L} \int_o^L \int_o^t Q(x,t') dx dt' \ .$$

Let $Q_i : = Q_r(0)$, i.e. the initial charge. It is now convenient to introduce the transfer rate ε:

$$\varepsilon(t): = Q_r(t)/Q_i \ .$$

The transfer efficiency may be defined by $\eta: = 1-\varepsilon$. If there was only self-induced transfer we would get

$$\varepsilon(t) = (1+t/\tau_s)^{-1} \ : \ \tau_s = \frac{2 \ L^2 \ C_{ox}}{\pi \mu \ Q_i}$$

If there was only diffusion we would get

$$\varepsilon(t) = \exp(-t/\tau_d) \ : \ \tau_d = 4 \ qL^2/\pi^2 \ \mu kT \ .$$

If there was only externally induced transfer we would get

$$\varepsilon(t) = \exp(-t/\tau_e) \ : \ \tau_e = 4 \ L/\pi^2 \ \mu E \ .$$

From the time scales we can see that for realistic cases self-induced transfer dominates initially. However, once that $Q_r(t)/C_{ox} \ll kTq$, diffusion takes over. Thus, diffusion is an important mechanism to transfer the charge package completely. (Externally induced transfer only gives a minor contribution in practice.) Transfer efficiencies may thus reach $\sim 10^{-4}$.

Let us now briefly mention some disturbing effects. There may be traps at the Si/SiO_2 interface, or in the bulk; they are characterized by the emission time,

$$\tau_{em} = \left[\sigma_n \, v \, n_i \, \exp \frac{E_t - E_F}{kT} \right]^{-1} \, ,$$

where σ_n is the cross section, E_t the energy of the trapping level, v the average thermal velocity, n_i the intrinsic concentration, and E_F the Fermi energy.

Noise may be generated by input, output, and transfer; for these sources we have

$$<\Delta n^2> \approx kTC/q \quad .$$

Furthermore the traps produce noise.

The most important problem is the dark current. Because of the external voltage, electron-hole pairs are generated spontaneously. It is this mechanism which limits the storage time (in practice to $\sim 1s$).

Let us now briefly present two possible mechanisms for writing into and reading from a CCD shift register. The procedures are immediately clear from the following figure.

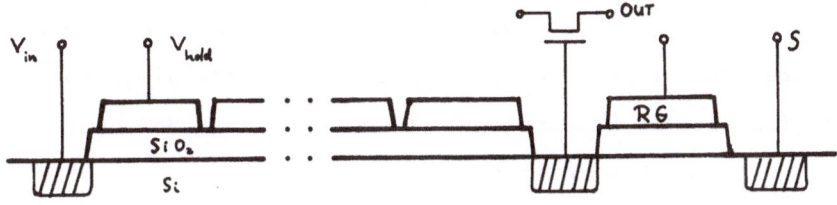

Fig. 4. Input into (left) and Output from (right)

a CCD Shift Register.

RG: reset gate, S: voltage supply.

The charge package injected into the first cell is simply given by
$Q = C (V_{in} - V_{hold})$. The charge package arriving at the last cell must
be sufficiently big to switch a transistor; it must therefore be at
least $5 \cdot 10^4$ elementary charges. (This corresponds to an areal charge
density of 10^{-12} cm^{-2}, if we assume that the size of a cell is about
$(2\mu m)^2$. The average distance between two elementary charges would be
about 100 Å or (in Si) 20 atoms.) In practice the logical value "1"
is represented by a cell filled to 80% of its capacity, the value "0"
by a cell filled to 20%. (The charge handling capacity of a CCD
shift register is an important restriction only for analog applications; we may ignore it here.)

Frequently several CCD shift registers are grouped together, as in
the SPS (Series-Parallel-Series) register, in order to reduce the
power dissipation. For this and other forms of organization we refer
to the literature.

III.   THE SUPERLATTICE SHIFT REGISTER

On a long term, planar technology will probably not be able to meet
the requirements of parallel processing /4/. Three-dimensional
technologies have to be developed. They admit more and shorter
connections and much more local memory.

Molecular Beam Epitaxy (MBE) may make part of such a technology /5/.
As a possible application we have suggested in a previous paper /6/
to use MBE grown doping superlattices as shift registers. We shall
briefly sketch the idea and then discuss the charge transfer
mechanism in more detail.

A doping superlattice is a periodic variation of the doping concentrations in an otherwise homogeneous and monocrystalline sample. So far
only one-dimensional superlattices have been made. The superlattice
period is typically in the range 100 ... 1000 Å.

A doping superlattice is said to be compensated if on the average the
numbers of donor and acceptor atoms are equal. In the ground state of
such a compensated doping superlattice all donor levels are empty
and all acceptor levels are occupied (apart from thermal excitation).
We have thus a periodic variation of negative and positive charges
which causes a periodic variation in the band edges.

The following example shows a compensated superlattice with a
constant doping concentration within the p resp. n regions which
generates a band edge variation of parabolic shape.

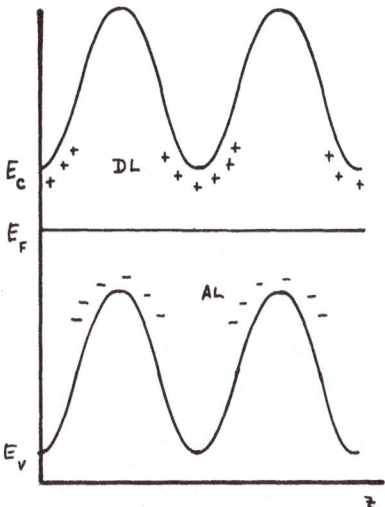

Fig. 5. Band Edge Variation in a Compensated

Doping Superlattice.

AL: acceptor levels, DL: donor levels

Such doping superlattices show many interesting features /7/. For our purposes the most interesting properties are:

1.   Charges may be stored in the valleys of the band edges.

2.   Because of the small dimensions, the motion in z direction will be quantized, and the allowed energies may be written

$$E_{tot}^{(s)} (\vec{k}) = E^{(s)}(k_z) + \frac{\hbar k_x^2}{2 m_x^*} + \frac{\hbar k_y^2}{2 m_y^*} \; ;$$

s labels the states according to the z quantization. (In a compensated superlattice $E^{(s)}(k_z)$ may simply be calculated from the Schrödinger equation with the band edge taken as the potential.)

3.   Due to the spatial separation of electrons and holes, non-equilibrium states with charges stored in the valleys have

an extremely long lifetime; lifetimes in the range 10^{-3} ... 1 s
may easily be achieved.

In the paper /6/ mentioned above we have shown that in a superlattice
tower of an area of (5μm x 5μm) and a height of 8μm one may store
100 bits of information.

We now come to the shift mechanism. Since we cannot think of any
technique which would allow for making "gates" to control the charge
transfer in such a shift register tower, a dynamic mechanism is required. This could be achieved in the following way. Each cell consists
of two valleys each of which contains two quantized states; these are
labelled A_n, B_n, C_n, D_n, where n refers to the cell number.

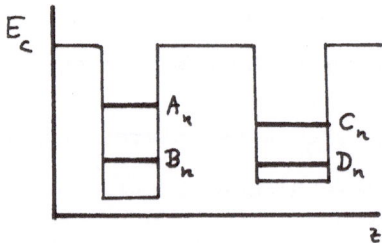

Fig. 6.   Superlattices Shift Register Cell

A charge package which initially is contained in the state D_n is then
transferred to the next cell in two steps. In the sample step, an
external field is applied such that D_n is in resonance with A_{n+1}. The
charge package will then tunnel into the state A_{n+1} and thermalize into
B_{n+1} which is a metastable state. In the hold step, the field is
lowered such that now B_{n+1} is in the resonance with C_{n+1}. Again, the
charge package will tunnel into C_{n+1} and thermalize into D_{n+1}.

Intuitively we might call the periodic variation of the external
field to shift the information the "milking" of the superlattice.
First estimates /6/ have shown that this mechanism should work at
77 K (nitrogen boiling point) with reasonable figures for cell sizes,
level splittings, transfer times, and power losses. However, electron-
phonon and electron-impurity scattering had been discarded. We shall
now discuss these effects. For a proper functioning of the shift
register we have to require

$$\frac{\Gamma_{resonant}}{\Gamma_{nonresonant}} > 10^4$$

where Γ denotes the transfer rate. This ensures that during a shift
cycle of ~200 transfers no information is lost. In the following
figure we consider the case that the levels of two adjacent valleys
are not in resonance.

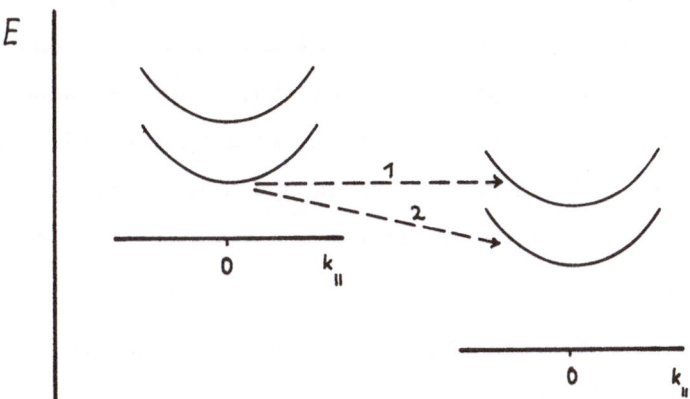

Fig. 7. Dispersions in Two Adjacent Valleys
(Nonresonant case); Transitions:
1: elastic scattering, 2: inelastic scattering

If there was no scattering a transition from one valley to the next would violate energy-momentum conservation. Hence, only scattering induced hopping is possible. We may distinguish between two processes: first, elastic scattering off an impurity, tunnelling to the next valley, and thermal relaxation; second, inelastic scattering off a phonon, causing a direct transition to the lowest level of the next valley.

Both processes have been studied in the literature in first order perturbation theory /8/ in the early 70's; at that time, however, people were mainly interested in negative differential conductivity mechanisms for superlattices.

In both cases the ratio of transfer rates is approximately given by

$$\frac{\Gamma_{resonant}}{\Gamma_{nonresonant}} \simeq (\Delta E)^2 \tau^2 / \hbar^2$$

where ΔE is the detuning between the levels (which is supposed to be bigger than the thermal energy kT). The increase proportional to $(\Delta E)^2$ is due to the fact that by detuning the electron states become more and more localized, and the overlap of the wave functions of neighbouring valleys decreases. (Strictly speaking, under resonance conditions first order perturbation theory is no longer valid; but we may use the expression given above as a rough approximation.) τ is the relaxation time of the scattering mechanism and may be determined from the scattering potential.

From the last two formulae we conclude that we must require

$$\Delta E > 10^2 \cdot \hbar/\tau .$$

If we insert for τ 10^{-13} s, which is the thermal relaxation time in the bulk at 77K, we find $\Delta E > 660$ meV, which is an unreasonable large

value (the band gap of Si is 1.12 eV). However, because of the
smaller phase space, relaxation times in a superlattice are larger
than in the bulk, and a more reasonable value for τ should be of the
order of $2 \cdot 10^{-12}$ s, thus reducing ΔE to ~33 meV. This is a value which
may easily be achieved. (A more careful calculation of τ does not help
much because of the limited validity of first order perturbation
theory.)

One may still improve the situation by confining the electrons to
regions without impurities ("modulation doping"), thus reducing the
probability for elastic scattering.

It should be noted that optical phonons only contribute to the
hopping mechanism if the detuning is of the order of multiples of
the optical phonon energies (~30 meV for GaAs and ~50 meV for Si).
Possibly one could use optical phonons to accelerate the transfer in
the resonant case.

An upper limit for the level splitting may be deduced from the band
gap. In order to avoid unwanted effects, like pair creation in the
external field or tunnelling into the continuum, the level splittings
should be smaller than one eighth of the gap energy (~140meV for Si).
The following figure shows the level splitting as a function of the
doping concentration; constant concentration is assumed, resulting in
a parabolic band edge (harmonic oscillator potential).

We conclude that a 77 K a doping concentration of $10^{18} \ldots 10^{19}$ cm^{-3}
is sufficient. For a superlattice to work at 300 K, we would need a
doping concentration of $10^{19} \ldots 10^{20}$ cm^{-3}; however, in this case we
would get level splittings which are already close to $^1/_8$ of the gap

energy of Si. Thus, it might be doubted that superlattice shift
registers would work at room temperature (unless one uses materials
with a larger band gap).

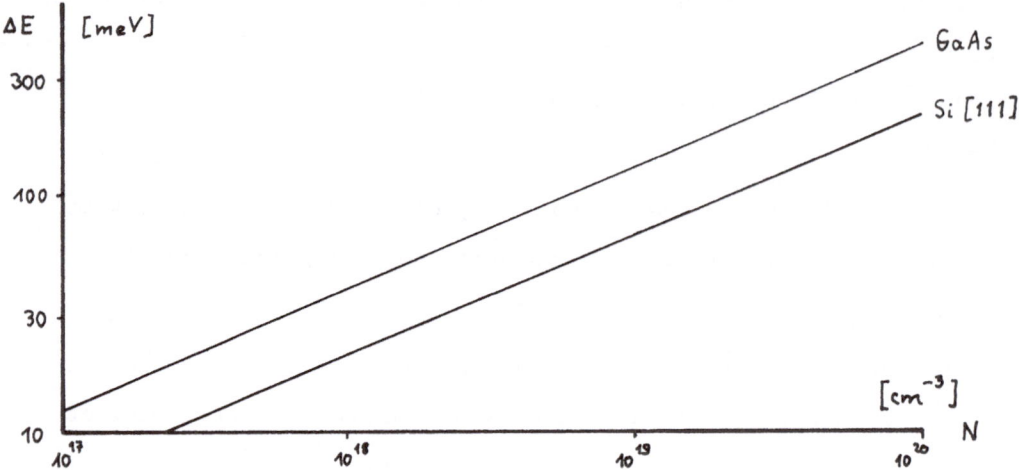

Fig. 8. Level Splitting as a Function of the Doping
Concentration.

For comparison: k·77K = 7 meV; k·300K = 27 meV)

At such high doping concentrations, the lowest energy level of the
valleys is already quite far away from the donor levels. Thus, our
discussion is still somewhat academic since the lowest energy states
are those of the donors at the centre of the valleys. Therefore, the
next step to be done is to reconsider the transfer process, including
the donor levels in the game.

IV. CONCLUSIONS

Parallel processing requires first of all to have the information
where it is to be processed. Therefore, local memories are necessary
for each processing element . The shift register is the most space-
saving type of memory.

In the first part of this talk we have reviewed the charge transfer
mechanism in CCD shift registers.

In the second part we have discussed the charge transfer mechanism in
a superlattice shift register. Such a superlattice would permit to
store a large amount of information in a very small volume. It has
been argued that electron-phonon and electron-impurity scattering (in
first order perturbation theory) should not cause an unbearable loss
of information during a shift cycle.

For a more realistic treatment, the impurity bound states will have
to be included.

ACKNOWLEDGEMENT

It is a pleasure to thank Drs. G.H. Döhler and R. Ruden for helpful
and stimulating discussions on the scattering mechanisms, as well as
Prof. I. Eisele, Dr. P. Vitanov, and Mr. A. Beck for discussions
concerning technological questions.

REFERENCES

/1/   S.D. Fouse, G.R. Nudd, A.D. Cumming, A VLSI Architecture for
      Pattern Recognition Using Residue Arithmetic. In: Proc. 6th
      International Conference on Pattern Recognition. IEEE Computer
      Society Press, 1982, p. 262 ff.

/2/   J. Becker, L. Cohen, G. Colella, A. Negro, G. Rispoli,
      G. Simoncelli, R. Tagliaferri, S. Tagliaferri, P. Zozzaro,
      Discussions on Artificial Intelligence, Parallel Computation,
      the Necessary Technology, and the Principle of Imperfection.
      Unpublished report (Italian/German), Salerno/Munich 1982.

/3/   L.J.M. Esser, F.L.J. Sangster, Charge Transfer Devices.
      In: C. Hilsum (Ed.), Vol. 4, of: T.S. Moss (Ed.), Handbook
      on Semiconductors. North-Holland, 1981.

/4/   J. Becker, I. Eisele, Computing with Neural Nets: Design and
      Technology. In: G. Musso (Ed.), Proc. Italian IAPR International
      Workshop on Cybernetic Systems. Salerno 1981; and I. Eisele,
      this volume.

/5/   I. Eisele, this volume.

/6/   J. Becker, Some Remarks Concerning the Use of Superlattices as
      Shift Register. In: S. Levialdi (Ed.), Image Processing from
      Computation to Integration. Academic Press. In press.

/7/   G.H. Döhler, K. Ploog, Periodic Doping Structure in GaAs.
      Prog. Crystal Charact. $\underline{2}$, 145 (1979).

/8/   R.F. Kazarinov, R.A. Suris, Electric and Electromagnetic Proper-
      ties of Semiconductors with a Superlattice. Sov. Phys. Semicond.
      $\underline{6}$, 120 (1972);
      G.H. Döhler, Electron States in Crystals with "nipi-Superstructure".
      phys. stat. sol. (b) $\underline{52}$, 79 (1972); G.H. Döhler, R. Tsu, L. Esaki,
      A New Mechanism for Negative Differential Conductivity in Super-
      lattices. Solid State Comm. $\underline{17}$, 317 (1975); and
      R. Tsu, G.H. Döhler, Hopping Conduction in a Superlattice.
      Phys.Rev. $\underline{B12}$, 680 (1975).

INTERPRETATION AND TIME

G. QUADRANTI

57, Bd. du Pont D'Arve, CH - 1205 Genève

A.   THE SITUATION

We want to find the first elements of the description of a subject
knowing reality.

What we want to describe is the knowledge that a subject can have.
More precisely we look for the terms which define the concept of
knowledge.

We have a constructive conception of knowledge. This means that we
try to describe knowledge with the distinction between two aspects.
One aspect is what we have to consider to be independent of the
subject's activity.

And the other is what we have to introduce with the subject's
activity. But we will have mixed situations, because the activity
can be exerted on constructed terms. And there are terms which are
not possible without the subject's activity.

To introduce this activity we begin with heuristic considerations.
This means we consider the situation which can be the "referent"
(i.e. the reality) of the theory. This situation is that of a small
child at the beginning. But we don't want to do a genetic research,
which analyses the formation of a concept by a child / 1 /.

If our aim was genetic theory we would not consider only a child
alone, but also the relation between the psychologist and the child.

In our theory we will have a theoretician, who writes a text, and a
subject (the phsychologist) described by this text as the psychologist
himself can describe a child. Both the theoretician and the psycho-

logist have to say something about the knowledge a subject has.

But the theoretician doesn't assert that the subject is a realisation of his theory on the ground of the behaviour of the subject. He constructs what he can and must later intervene in a description whose semantics can be justified by a perception of a behaviour. In this sense the theoretician does not perform a genetic research. His research is a priori and the semantics is asserted without perception. In other words, the theoretician builds his theory before any experiment is performed, but any empirical theory (e.g. genetic theory) must include this theory.

The a priori aspect will be justified if the theory can be considered as universal.

The genetic research may be introduced later when the theoretician can describe a subject (the psychologist) in a world, which recognizes in this world a behaviour (the child's behaviour). Then the theoretician can say how the subject (the psychologist) interprets the behaviour. That is not our aim. But with this remark we can introduce our main notion of interpretation.

When the theoretician describes a subject which distinguishes a behaviour and interprets it, then this subject is the most simple psychologist or, more generally, this subject finds himself in the most simple situation of communication.

Which are the most important elements of this situation? Essentially such a subject has to distinguish the behaviour (which he perceives), from what he asserts from this perception (but doesn't perceive). We will call Interpretation what will be asserted in this way. Now, if the theoretician wishes to describe this situation, he has to distinguish in his language two subsets of symbols, one whose realisation will be "the perception of the subject" (the psychologist) and the other whose realisation will be "the interpretation" which the subject asserts.

Moreover the theoretician can describe the beginning of a genetic research. Such a research is not our intention.

But we introduce the Assumption that this structure is universal.

This means the theoretician needs this decomposition of his language (into two subsets of symbols with relations) not only for the description of the subject's interpretative activity of a behaviour but also for the description of his knowledge of the world.

This means we consider physics as an interpretative construction. Why?

> Because we admit that time plays an important role.
> But for us time is not (or not only) a perception.
> And physics (this means the world) has to be constructed over time.

We are now able to state precisely our aim:
to construct a notion of time with the concept of interpretation.
(The formulation will not be strongly formalized.)

Before we start we have to state precisely on one hand, what the theoretician can write and on the other how he can introduce semantical rules that means how to give the rules that allow to say when something corresponds to what he wrote. But, as we have said, the theoretician doesn't find reference (reality) for his theory with the perception but the theoretician asserts that his theory describes a subject. We will say he attributes something to the subject.

Then to fix semantical rules is equivalent to fix rules for the attribution. This terminology allows us to distinguish the interpretation, which is in the language of the theoretician, from the Attribution (semantics).

B.   PERCEPTION AS A FIRST LEVEL OF REALITY

We will introduce now the first set of symbols for the theoretician. With this set of symbols he can describe the perception of a subject. We will have two subsets as symbols of this first set.

We consider the following symbols

P $= \{p_1 \ldots p_n\}$ p describes one possiblity of perception

E $= \{a,b\}$ a describes the presence
b the absence

E_p $= \{p\} \times E$ this allows to express the presence or absence of perception

D $= \prod_p E_p$

An element of D expresses the fact that an element of P is perceived (absent or present) together with the presence or absence of other elements. D stands for the French word "donné"; we may call it the "data" set.

The theoretician can write any element of D. Such an element expresses a combination of perceptions.

To write an element of D, the theoretician can use the symbols (,). We won't talk now about "attribution".

We can remark the following:

1) With D we have admitted already that p, a or b can be repeated, this means they can be recognized.

2) Moreover with D we admit simultaneity, but not an order relation.

The second subset will be the set of symbols which allows to express the "consideration".

We distinguish between"to have a perception" and "consider a perception".

The subset can interpret perceptions as far as he considers them. The theoretician can introduce symbols for sets he constructs in the usual manner over D.

D and symbols of sets are the first set of symbols.

C.   THE INTERPRETATION OF THE PERCEPTION AS A SECOND LEVEL OF REALITY

We will now introduce the second set of symbols. These symbols allow to express the interpretation.

An interpretation will not be considered as perceived.

What can and must the theoretician write if he wants to describe one act of interpretation?

He has to introduce a set like those constructed over D. Let M be the symbol of this set.

Then he introduces a new symbol for a set, let it be I.

We have to fix conditions about M and I. M stands for "message" and I for "interpretation".

1) $M \cap I = \phi$ (empty)

2) The theoretician considers I as the set of interpretations of M. This means the theoretician considers a function f from M to I.

f:M → I (f is surjective)

this means /M/ ≥ /I/.

3) The theoretician must identify every function f' he obtains from f only by permutation of the images (in I)

$$\left\{ \begin{array}{l} \text{if } f(m) = f(m') \text{ then } f'(m) = f'(m') \\ \text{if } f(m) \neq f(m') \text{ then } f'(m) \neq f'(m') \end{array} \right\} \iff f' \equiv f$$

4) M and I can be sets of sets. But in this case we ask for the elements of M to be disjoint.

5) We ask for f to be a one-to-one function.

It is an open question if these conditions are sufficient to define a general theory of interpretation.

D. A FUNDAMENTAL INTERPRETATION: TIME

Time allows the definition of an order relation.

We are now able to construct the notion of time. We must only define the set M.

In a first approximation M can be a set of perceptions, this means that M is a subset of D. But this is not sufficient. We need more general considerations. We have to state precisely our intuitive notion of time. What do we understand by time?

First, an order relation.

Second, the possibility of repetition of a perception.

Here we consider together "concepts" (order relation) and "reality" (perception). But this is inevitable because the theoretician must describe both the concept of a subject and his reality. (If the concepts correspond to the syntax we have a mixture of syntax and semantics).

Until now the reality for the subject can be expressed by elements of D. This means that these elements must satisfy the order relation. But these elements must be repeatable.

Let us consider this notion.

The theoretician can attribute many elements of D. He can write many times the same symbol of an element of D, he can also write a family of elements of D. But this doesn't mean that for the subject the elements of D are repeated. We can say that now the subject doesn't have the capacity to distinguish two different occurrences of the same element of D.

(In a completely formalized theory this remark must appear in the rules of attribution. This means the theoretician can not use a family to attribute the repetition. He must construct a new relation for this).

CONCLUSION

If the theoretician disposes only of D and its subsets, he can't express that for the subject there is repetition. He needs a new para-meter. It is impossible to use order relation over D to express the repetition.

We can express the repetition with the order relation only by d < d, where d is an element of D. But this is impossible.

We can admit a special set of perception, with order relation, which is the new parameter. That is the traditional solution. And we will have such a set, but it won't be a perception, otherwise we have only changed the definition of P.

The problem is ill defined.

We must start with the notion of repeatability and not with the order relation.

The definition of the set D doesn't allow repetition of any of its elements. The theoretician may write the same element twice but the psychologist has no means to distinguish them. To distinguish them we have to introduce a new parameter. If we consider this parameter as perceived we change the definition of D and P. Hence this parameter cannot be perceived. Thus we will use the interpretation. The repeatability will allow us to define an order relation.

We will construct the repeatability.

1) The theoretician can write every subset δ of D.

2) He can introduce new symbols for sets (which we will define).
    Let them be M; I; Δ; f. δ_0, δ_2 ...
    and symbols for elements

    d_1...   elements of D or δ (data)

    $\tau_{0,1}$...   elements of I (interpretations)

    m_1...   elements of M (messages)

    M,I are countable

    δ_0, δ_1... are subsets of D.

We ask for the following conditions.

3) M, I, Δ and f are not empty.

4) $I \cap D = \emptyset$

5) There is a unique δ_o, $\delta_o \varepsilon M$

   M is the set that the subject has to interpret.

   I is the set of interpretations of M.

   $f:M \to I$; f is a one-to-one function and $\tau_o = f(\delta_o)$

6) Δ is a function from I to P'(D), where P'(D) = P(D)\Ø

   this means the set of δ_o, $\delta_1 \ldots$

   Δ is a new function which we have to introduce as hypothesis as D.

   Δ permits us to fix the order relation!

7) finally

   $M = \{\delta_o\} \cup \Delta$

   We can represent graphically the situation as follows:

$$\delta_o \longrightarrow f(\delta_o) \quad = \tau_o$$

$$\delta_1 = \Delta(\tau_o) \ (\tau_o, \delta_1) \longrightarrow f(\tau_o, \delta_1) = \tau_1$$

$$\delta_2 = \Delta(\tau_1) \ (\tau_1, \delta_2) \longrightarrow f(\tau_1, \delta_2) = \tau_2$$

subsets of D	M	I
(data)	(messages)	(interpretations)

Δ permits the repeatability and I is the new parameter.

E. TIME ALLOWS THE DEFINITION OF AN ORDER RELATION

We can now define an order relation.

1) First we introduce the notion of a neighbour.

Two elements of M, (τ,δ) and (τ',δ') are neighbours <u>if and only if</u> we have $f(\tau,\delta) = (\tau', \delta') = \tau$.

f is a one-to-one function, this implies that every element different from δ_0 has a neighbour and at most two neighbours.

2) We can define a set of neighbours.

This is a subset V of M every element of which has two neighbours except two which have only one neighbour.

We consider $V = \{m\}$ as a set of neighbours.

3) Set of neighbours of m; V_m

V_m is a set of neighbours which does not contain m but contains a neighbour of m.

4) The past of m; H_m that is a set of neighbours of m, V_m, which contains δ_0.

5) Order relation

We say that $m > m'$ only if $H_m \supset H_{m'}$.

In conclusion we have constructed a clock for one observer.

F. OUTLOOK

1) The next problem we shall try to deal with is the construction of space using the concept of time. Then we shall try and define motion and acceleration.

This will lead to the concept of mass as interpretation of the acceleration.

2) Furthermore we shall deal with the notion of causality, of which a possible definition could be: "Causality is the identity relation over the interpretations".

3) Given a good theory of causality the next steps should be the foundations of quantum mechanics and special relativity.

REFERENCE

/ 1 / J. Piaget, Introduction à l'épistémologie génétique. P.U.F. 1973

A STOCHASTIC MODEL OF 1/f NOISE AND ITS APPLICATION TO SEMICONDUCTORS

F. Grüneis[+)]
Friedrich-Herschelstr. 4
D-8000 München 80

1.Introduction

1/f noise has been observed in many physical, technical and biological systems (for a survey see Wolf /1/).For semiconductors, an empirical relation had been found by Hooge /2/

$$S(f) = \bar{I}^2 \, \alpha_H / (\bar{N}_{TOT} f) \qquad (1.1)$$

whereby

$\bar{I}$ = the mean value of the electric current
$\bar{N}_{TOT}$ = the mean number of charge carriers per volume
α_H = a numerical constant
f   = frequency
S   = power spectral density (psd)

Relating to equation (1.1), the following questions arise:

- is α_H a universal or a material constant ?
- what is the mechanism of 1/f noise ?

Hooge /2/ found that for semiconductors $\alpha_H \sim 2 \times 10^{-3}$. Since
$\alpha_H \sim 10^{-6}$ in silicon wafers /3,4/ , α_H cannot be a universal
constant.

Though several mechanisms have been proposed to explain 1/f noise
/5,6,7/, some features are open to question.

Recently, a stochastic model of 1/f noise has been proposed /8/ , which
is based on a doubly stochastic Poisson process (DSPP). The results
of this theory are summarised in the next chapter. In chapter 3,
this model is applicated to 1/f noise in semiconductors.

+)Consultant to

  Fraunhofer-Institut für Hydroakustik
  Waldparkstr. 41
  D-8012 Ottobrunn b. München

2. A number fluctuation model of 1/f noise

In Fig.1 the time signal of a special DSPP is shown.

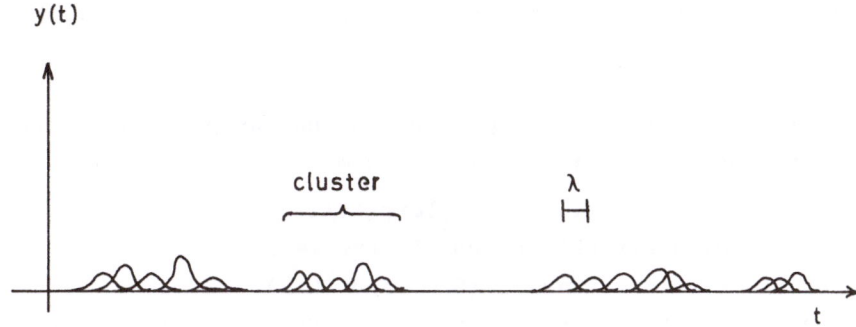

Fig. 1 Time signal of a special DSPP

This DSPP consists of a succession of clusters, having the following features:

- the time points of the occurrence of the clusters follow a
  Poisson process

- the mean number of clusters is $\bar{n}_c$.
  A cluster is a time limited Poisson process, having the following
  features:

- the mean time between two events within a cluster is $\bar{\lambda}$

- the probability of finding exactly N events within a cluster is P(N)

- N can take on values within range 1,2,..., N_O;
  $\bar{N}$ is mean value and $\overline{N^2}$ second moment of N.

The psd of this DSPP has generally been dealt with by Baiter,
Grüneis and Tilmann /9/. The shape of the psd of this DSPP is strongly influenced by the distribution P(N). In principle, any arbitrary
choice can be made for the distribution P(N), giving rise to very
different shapes of the psd.

But the distribution

$$P(N) = N^{-2} / \sum_{N=1}^{N_o} N^{-2} \qquad (2.1)$$

is the only one giving rise to a 1/f pattern. At the same time, this distribution has the extraordinary property, that the relative variance of such a DSPP is a maximum compared to all other distributions P(N) /8/.

In the following, the case is considered that an electric current I(t) is generated by single events of the form

$$x(t) = (q/\bar{\tau}_s)\exp(-t/\bar{\tau}_s) \qquad (2.2)$$

q being the elementary charge and $\bar{\tau}_s$ the mean lifetime of the charge carriers. Taking (2.1) and (2.2) into account, the psd of such a DSPP is given by (equation (5.3) of /8/):

$$S(f) = \left[2q\bar{I}+\bar{I}^2\alpha_H/\bar{N}_{TOT}f) \right] / \left[1+(2\pi f\bar{\tau}_s)^2\right]+\bar{I}^2\delta(f) \qquad (2.3)$$

whereby

$\bar{I} = \bar{n}_d q$ = the electric current and
$\bar{n}_d = \bar{n}_c\bar{N}$ = mean number of charge carriers of the DSPP. As a result of this theory, α_H has been found to be (equation (4.8) of /8/)

$$\alpha_H = 2\bar{\tau}f_1 \qquad (2.4)$$

In this equation $\bar{\tau}$ can be interpreted as transit time, which is defined by

$$\bar{N}_{TOT} = \bar{n}_d\bar{\tau} \qquad (2.5)$$

In semiconductors $\bar{\tau}_s$ and $\bar{\tau}$ are not identical, as it is the case in vacuum tubes or ionic solutions for example. In semiconductors

$$\bar{\tau} \lesssim \bar{\tau}_s \qquad (2.6)$$

in other words, it is supposed that the charge carriers do not recombine on their way through the semiconductor material.
In equation (2.4), f_1 is the upper cutoff frequency of 1/f noise. Beyond f_1, 1/f noise is drowned in shot noise, the psd of which is $2q\bar{I}$ (see (2.3)). f_1 is given by (equation (3.9) of /8/):

$$f_1 \cong 1/(3\,\bar{\tau}_c) \qquad (2.7)$$

$\bar{\tau}_c = \bar{\lambda}\bar{N}$ being the mean duration of the clusters.
The lower cutoff frequency f_o of 1/f noise is related to $\bar{\tau}_o \cong \bar{N^2\bar{\lambda}}$ which is the mean time of the longest clusters, by

$$f_0 \cong 1/(3\bar{\tau}_0) \tag{2.8}$$

Below f_0, $S(f)$ is constant.

Thus the range of 1/f noise is given by

$$f_1/f_0 \cong \overline{N^2}/\bar{N} \cong N_0/(\ln N_0 + C_E) \tag{2.9}$$

and therewith depends only on N_0, the maximum value of N of the distribution (2.1). ($C_E = 0.5772$ is Eulers constant).

It has already been remarked, that the psd is strongly dependent on P(N). Deviations from P(N) of (2.1) which can be written in the form

$$P(N) = N^Z / \sum_{N=1}^{N_0} N^Z \tag{2.10}$$

give rise to a psd of the form $1/(f\bar{\lambda})^{b(z)}$ with $0 \leqq b(z) \leqq 2$. For $z = -2$ one has (2.3) again. Deviations from a 1/f shape can be explained in this way.

3. Physical interpretation of a DSPP and its application to the physics of semiconductors

The question arises, how this stochastic model fits together with the physics of semiconductors. Two points have to be observed thereby, the formation of clusters of charge carriers and the distribution P(N) of the number of charge carriers per cluster.

Regarding the first point, the formation of clusters of charge carriers can readily be imagined under the supposition that there are emitting centers which emit charge carriers in clusters. On the basis of this model the emission rate of these centers should be $1/\bar{\lambda}$. Regarding a semiconductor device, these emitting centers might be found in the contacts of the semiconductor material. Under the supposition that this condition is to be met, one can say, that 1/f noise is originated by the current itself. In this case, there is no need introducting mobility fluctuations to explain 1/f noise in semiconductors /6/.Regarding the second point, concerning P(N), there is no change of a direct observation of single events because of the overlap of the fluctuating events. Thus, indirect methods have to be used. This can be done in the following ways:

1/f noise has been observed over a minimum range of 10 decades /1/; thus

$$f_1/f_0 \geqq 10^{10} .$$

From (2.9) one can calculate $N_o \geqq 10^{12}$.

In semiconductors, the upper frequency f_1 of 1/f noise is in the range of 10^4Hz /1/.

From (2.7) $f_1 \cong 1/ (2\bar{\lambda}lnN_o)$ one has $\bar{\lambda} \cong 10^{-6}$ sec^{-1}. Thus, the emission rate of the emitting centers turns out to be $1/\bar{\lambda} \sim 10^6$ sec. Unfortunately, experimental values are missing for this quantity.

On the other hand, equation (2.4) gives a relation between α_H, the transit time $\bar{\tau}$ and f_1, the upper cutoff frequency of 1/f noise. Taking the experimental values of /4/ for $f_1 \sim 10^3$Hz and $\bar{\tau} \sim 0,5 \times 10^{-9}$ sec, α_H turns out to be $\alpha_H \sim 10^{-6}$ in good agreement with the experimentally found values.

It should be remarked that the transit time can vary considerably for individual wafers, because it is dependent on the geometry of the device. This might be an explanation for the wide spreading of α_H for a certain type of semiconductor material.

4. Summary of results and conclusion

1/f noise can be explained on the basis of a DSPP /8/ . Applying this stochastic model to semiconductors, Hooges constant α_H is found to be a material constant which can take on arbitrary values. On the basis of this model there have to be postulated emitting centers, which could be found in the contacts of the semiconductor material. These emitting centers emit charge carriers in clusters. This cluster formation gives rise to 1/f noise. This stochastic model can be tested by a comparison of experimental values of α_H with the theoretical value of $\alpha_H = 2 \bar{\tau} f_1$, $\bar{\tau}$ being the transit time and f_1 the upper cutoff frequency of 1/f noise.

References

1) D. Wolf, Noise in Physical Systems, Proceedings of the Fifth
   International Conference on Noise, Bad Nauheim, March 13-16 (1978)
   122, Springer-Verlag, Berlin, Heidelberg New York

2) F.N. Hooge, Physics Letters 29A,3 (1969) 139

3) R.D. Black M.B. Weissman, P.J. Restle
   Journal of Applied Physics 53,9 (1982) 6280

4) J. Kilmer, A. van der Ziel, G. Bosman,
   Solid-State Electronics 26,1 (1983) 71

5) A. McWorther, Ph.D. thesis, Massachusetts
   Institute of Technology, Cambridge, Massachusetts (1955)

6) F.N. Hooge, T.G.M. Kleinpenning, L.K.J. Vandamme,
   Rep. Prog. Phys. 44(1981) 479

7) R.F. Voss, J. Clarke, Physical Review B 13,2 (1976)556

8) F. Grüneis, PHYSICA A in press

9) H.-J. Baiter, F. Grüneis, P. Tilmann,
   International Symposium on Cavitation Noise, presented at the
   Winter Annual Meeting, ASME, Phoenix, Arizona, November 14-19
   (1982)93

NON DETERMINISTIC MACHINES AND THEIR GENERALIZATIONS
A. Bertoni - G. Mauri - N. Sabadini
Istituto di Cibernetica - Università di Milano

1. INTRODUCTION

There are some slightly different senses in which the words "non determinism" (and "non deterministic") have been used in computer science and, in a more general context, in system theory; hence, a full understanding of non determinism can be reached not only on the basis of a rigorous formal definition, but also by studying the effects it has on the different aspects of the systems behaviour, so that it is possible to grasp all the different nuances of meaning of non determinism and the subtle differences among it and such related, but different, concepts as concurrency, randomness, parallelism, which are often confused with it.

Intuitively, a non deterministic system is a system for which we have only an incomplete knowledge of all of the factors which can influence its evolution. In the case of automata and Turing Machines, this fact is expressed, at a rather abstract level, as the possibility of reaching, from a given configuration A, a set of different configurations $B_1, \ldots, B_n$ in a single step.

It is well known that Turing Machines (TM's) were introduced as a formal model of an algorithm or computation, in order to exactly define the notion of computable function. Such a model is a deterministic one, in the sense that a given input generates a unique sequence of computation steps producing the output. Even if some other formalisms, proposed more or less in the same period with the same goal - such as rewriting systems - are implicitly non deterministic, non determinism was firstly introduced as an explicit feature of a computation model - the finite automaton - by Rabin and Scott [18], and then extended to other classes of machines, including TM's.

A first problem about non determinism, is whether it adds computational power to machines, in the sense that it allows to compute functions that cannot be computed by the corresponding deterministic machines. The most significant results in this direction show that finite automata and Turing Machines take no advantage from non determinism, while non deterministic pushdown automata, on the contrary, can compute functions, for example the characteristic function of the set ww^R, where w is a word on a finite alphabet Σ and w^R its reversal, which cannot be computed without non determinism.

However, it was the development of the theory of computational complexity, and in particular of NP-completeness, which pointed out the enormous conceptual importance of non determinism, since the study of the relationship between deterministic and non deterministic complexity classes gave a very deep insight into the whole topic of computational complexity and stimulated interesting developments in various directions, which we will discuss in the following.

The first important topic is the comparison between the power of non deterministic (ND) and deterministic (D) Turing Machines with given complexity bounds, in particular polynomial. While NDTM's have been proved to be equivalent, within a polynomial bound, to DTM's with respect to space complexity by Savitch [20], their equivalence, or non-equivalence, with respect to time complexity is still an open problem.

More recently, extensions of NDTM have been given in two directions. In [4] Chandra, Kozen and Stockmeyer introduce Alternating Turing Machines (ATM's), which generalize NDTM's, and compare them with DTM's. ATM's can be seen as abstract models of parallel computers [15], and the relations among deterministic and alternating complexity classes support the so called "parallel computation thesis" [10], which states that time in a parallel machine (with unbounded parallelism) is polynomially related to space on a serial machine. Furthermore, NDTM's have been compared with other models of machines such as Random Access Machines with arithmetical primitives [22] or Vector Machines [17], which implicitly implement a form of parallelism.

The second extension of NDTM's aims at covering counting problems as well. This generalization, from decision problems to enumeration problems, has been carried out by Simon and Valiant [23], who introduced the concept of Counting Turing Machine as a formal model for describing enumeration problems. The power of Random Access Machines in solving enumeration problems has been then studied by Bertoni, Mauri and Sabadini [3], giving a strong characterization of enumeration problems that can be expressed in polynomial space on Turing Machines.

2. NON DETERMINISTIC TURING MACHINES AND COMPLEXITY CLASSES

Complexity theory has been developed mainly with respect to decision problems. We can define a decision problem as a pair $<I,X>$, where I is a (numerable) set of instances and $X \subseteq I$: given $i \in I$, we have to decide whether $i \in X$. Usually, the set I is encoded by means of the set Σ^* of words over a finite alphabet Σ.

Let us now give the standard definition of Non Deterministic Turing Machine and see how we can solve decision problems by means of such machines:

Def.1 - A Non Deterministic Turing Machine (NDTM) is a seven-tuple
$$M = <Q, \Gamma, \Sigma, \delta, q_o, q_y, q_n>$$
where:
Q is the finite set of states;
Γ is the finite tape alphabet, containing a special symbol b called blank;
$\Sigma \subseteq \Gamma - \{b\}$ is the input alphabet;
$\delta : Q - \{q_y, q_n\} \times \Gamma \longrightarrow 2^{Q \times \Gamma \times \{L,R\}}$ is the next move function;

$q_o \in Q$ is the initial state;

$q_y \in Q$ is the accepting state;

$q_n \in Q$ is the rejecting state.

Def.2 - A configuration of a NDTM M is a string $vq\delta w$, where $v, w \in \Gamma^*$, $\delta \in \Gamma$ and $q \in Q$. A configuration $vq\gamma w$ will be called initial, accepting or rejecting if, respectively, $q=q_o$, $q=q_y$, $q=q_n$.

To explain the activity of a NDTM M, let us have a device consisting of a finite control, a (infinite) tape marked off into cells and a read-write head that scans the cells and sends information to the control. The configuration $vq\delta w$ will describe the situation in which the control is in the state q, the string $v\delta w$ is written on the tape (one symbol per cell, with blanks in all the remaining cells), and the symbol δ is scanned.

Def.3 - Let $A=v\delta'q\delta w$ be a configuration of a NDTM M, and let
$$\delta(q,\delta) = \{<q_1, \delta_1, d_1>, \ldots, <q_n, \delta_n, d_n>\}.$$
For every $<q_k, \delta_k, d_k>$ we can construct a configuration $B_k = v\delta'\delta_k q_k w$ if $d_k=R$ or $B_k = vq_k\delta'\delta_k w$ if $d_k=L$. We say that B_k derives from A, and write $A \longmapsto B_k$.

The meaning is that, if the machine configuration at time t of a discrete time scale is A, then the machine can enter a new configuration B_k at time t+1 by choosing a triple $\langle q_k, \delta_k, d_k \rangle \in \delta(q, \delta)$ and then changing the state from q to q_k, substituting δ_k for δ and moving the head one cell right or left.

Deterministic Turing Machines (DTM's) correspond to the particular case where every configuration can have at most one derived configuration.

Def.4 - Given a NDTM M and an input word $w \in \Sigma^*$, a computation of M is a sequence of configurations $q_o w = A_o \mapsto A_1 \mapsto \dots \mapsto A_n \mapsto \dots$

Hence, the machine starts in the initial state q_o with the word w written on the tape, and with the head positioned on the first symbol of w, and then performs its computation, until it eventually reaches an accepting or rejecting configuration. Obviously, in the deterministic case there is exactly one computation sequence for every input w; in the general non deterministic case, the different computation sequences which can be generated by an input w, can be arranged in a computation tree, in which the root is labelled by the configuration $q_o w$ and every node of label A has a son of label B_k for every $A \mapsto B_k$. We can now define the set accepted by a NDTM.

Def.5 - Let M be a NDTM and $w \in \Sigma^*$. M accepts w iff the computation tree of M has at least one accepting leaf. $L_M = \{w \in \Sigma^* / M \text{ accepts } w\}$ is the set accepted by M.

It has been proved that non determinism does not increase the "accepting power" of TM's, since for every set accepted by a NDTM there is a DTM accepting it. But now we can ask whether DTM's are as "efficient" as NDTM's in accepting sets, i.e. whether accepting a given set L requires the same amount of resources, in particular time and space, both on a DTM and a NDTM. Hence, we have to exactly define the notions of time and space consumption by TM's.

Def.6 - A NDTM M has:
- time complexity T(n) if for every $w \in \Sigma^*$ of length n the shortest accepting computation, if any, has length T(n) at most;
- space complexity S(n) if for every $w \in \Sigma^*$ of length n no sequence of computation requires more than S(n) tape cells.

These definitions, which are easily particularized to the case of DTM's, allow to classify sets on the basis of the complexity of the machines that accept them. Here, we will use the following form to represent in a uniform way the different complexity classes we will consider:

M-BOUND-RESOURCE

where M is a symbol representing the type of machines (D, often omitted, for deterministic, N for non deterministic and so on) and RESOURCE may be TIME (often omitted) or SPACE. Furthermore, we will group the bounds into the three classes LOG (logarithmic bounds), P (polynomial bounds) and EXP (exponential bounds).

Of particular significance is the class P of the sets accepted in polynomial time - w.r.t. the length of the input - by some DTM, since this class corresponds to problems that are considered as practically solvable on real machines [1,5] . If we consider NDTM's, we have the class NP of the sets accepted in polynomial time by some NDTM. The polynomial classes with regard to space complexity are PSPACE and NPSPACE respectively; furthermore, we have EXPTIME, EXPSPACE and so on. Finally, a particular discussion is needed in order to introduce the class LOGSPACE: in fact, with the above definition of TM's, the minimum space required is exactly the length of the input string. Hence, it is necessary to modify the definition, by introducing a working tape distinct from the input tape, and considering as space complexity the amount of working tape used [1,5] .

It is obvious that, for every bound and every resource, the inclusion

$$\text{D-BOUND-RES} \subseteq \text{N-BOUND-RES}$$

holds. In particular, we have:

$$P \subseteq NP \qquad \text{and} \qquad \text{PSPACE} \subseteq \text{NPSPACE}.$$

But, are these inclusions proper or, in other words, NDTM's are effectively more efficient than DTM's? As far as space complexity is concerned, the following result has been proved by Savitch [20] :

Theorem - Every NDTM M working in space S(n) can be simulated by a DTM M' working in space $S(n)^2$.

Hence, we have PSPACE=NPSPACE, and no substantial advantages are given by non determinism with respect to space consumption within a polynomial bound. On the other hand, it is still an open question whether NP contains some sets that cannot be accepted in polynomial time by any DTM, i.e. whether P=NP or not. The best general result we can state about the time needed to simulate a NDTM by a DTM is the following

Theorem - If $L \in NP$, then there exists a polynomial p such that L can be accepted by a DTM having time complexity $O(2^{p(n)})$. This implies that $NP \subseteq EXPTIME$.

The problem P=NP? , while not yet solved, has been one of the most stimulating questions in theoretical computer science, and has led to the development of a number of fundamental concepts such as polynomial reducibility among sets and NP-completeness.

Def.7 - A set $L_1 \subseteq \Sigma_1^*$ is said to be <u>polynomially reducible</u> to a set $L_2 \subseteq \Sigma_2^*$ (in symbols, $L_1 \xrightarrow{p} L_2$) iff there is a polynomial time DTM M which will transform every word w on Σ_1 into a word w' on Σ_2, so that $w \in L_1$ iff $w' \in L_2$.

Lemma - Let $L_1 \xrightarrow{p} L_2$; then if $L_2 \in P$, then $L_1 \in P$.

On the basis of this lemma, we can single out a very important subclass of NP, the class of NP-complete sets.

Def.8 - A set L is NP complete iff:
  a) $L \in NP$ and
  b) every set in NP can be polynomially reduced to L.

This means that, up to a polynomial time translation, to accept an NP-complete set is equivalent to accepting any other language in NP. Hence, a polynomial time DTM accepting any NP-complete set would allow to accept, in deterministic polynomial time, any other set in NP, thus proving that P=NP.

Some hundreds of sets are known to be NP-complete (for a list, see [3]), among which there are sets encoding some very significant problems in the areas of graph theory, network design, scheduling, information retrieval, mathematical programming, algebra and number theory, logic, automata and languages. We will quote here only the most well known of them, the satisfiability problem for formulas of propositional calculus, which was the first problem proved to be NP-complete by Cook [5]. This problem can be specified as follows:

SATISFIABILITY
INSTANCE: A formula F of propositional calculus in Conjuntive Normal Form.
QUESTION: Is there an assignment of truth values which satisfies F?

There are several possibilities for implementing the computations of a NDTM :

a)   first, we can transform them into deterministic computations by exploring
     successively all the paths of the computation tree, and going back to the
     nearest node with unesplored existing arcs, whenever a rejecting leaf is
     reached; this mechanism is known as backtracking and is obviously very
     expensive in time .

b)   we can create a new copy of the machine for each triple in $\delta(q,\gamma)$ and let all
     the copies work in parallel . Hence, non determinism is implemented by
     unbounded parallelism and we have an exponential growth of the number of
     copies.

c)   at the end, a way to bypass the nondeterminism is to give a criterium for
     choosing a unique triple in $\delta(q,\gamma)$. The non deterministic computation
     corresponds, in this way, to the possibility of having an "oracle" that
     "guesses" the correct strategy (if it exists).

     The probabilistic version of this point of view introduces various notions of
     random machines, which will not be discussed here.

3.ALTERNATING TURING MACHINES

     A generalization of NDTM has been introduced by Chandra, Kozen and Stockmeyer
[4] who studied the effects of a new capability, called alternation, on complexity
classes.

Def.9 - An <u>alternating</u> <u>Turing</u> <u>Machine</u> (ATM) is a NDTM whose set of states is
     partitioned into two disjoint subsets E (<u>existential</u> states) and U (<u>universal</u>
     states).

     A node labelled vqw in a computation tree of an ATM is <u>accepting</u> iff:
a) it is a leaf and q is an accepting state ; or
b) it is internal, $q \in U$ and all of its sons are accepting; or
c) it is internal, $q \in E$ and it has at least one accepting son.

     The computation tree generated by an input word w is accepting if its root is
an accepting node. In order to decide whether a computation tree is accepting, we
can associate the value 1 to accepting nodes and 0 to non accepting ones, starting
from leaves; the value for a node corresponding to a universal (resp. existential)
state is the boolean "and" ("or") of the values of its sons. Obviously, a NDTM is an
ATM with existential states only.
     For ATM, we must consider accepting subtrees rather than accepting paths:

Def.10 - An <u>accepting</u> <u>subtree</u> of an accepting tree T for an ATM M is a subtree T_1 of
     T such that:

     a) T_1 contains the root of T;

     b) if x is a universal node in T_1, then all of its sons are in T_1 and   are
        accepting;

     c) if x is an existential node in T_1, then exactly one accepting son of x is in
        T_1.

Def.11 - An ATM M has:
     - time complexity T(n) if every accepting tree with root $q_o w$, $|w|=n$, contains an
     accepting subtree of height T(n) at most;
     - space complexity S(n) if it uses no more than S(n) cells for the computation.

     We can now introduce the classes:

$$\text{ALOGSPACE} \subsetneq \text{APTIME} \subsetneq \text{APSPACE} \subseteq \text{AEXPTIME} \subseteq \text{AEXPSPACE}$$

and compare them with the corresponding classes for deterministic machines. In [4] the following results relating alternating and deterministic complexity classes are given.

Theorem - a) $NSPACE(S(n)) \subseteq ATIME(c.S(n)^2)$ for $S(n) \geqslant n$;

       b) $ATIME(T(n)) \subseteq DSPACE(T(n))$ for $T(n) \geqslant n$;

       c) $ASPACE(S(n)) \subseteq DTIME(c^{S(n)})$ for $S(n) \geqslant \log(n)$;

       d) $DTIME(T(n)) \subseteq ASPACE(c.\log T(n))$ for $T(n) \geqslant n$

       and hence

       e) $ASPACE(S(n)) = DTIME(c^{S(n)})$ for $S(n) \geqslant \log(n)$.

From the above results and the obvious inclusions the following equalities:

$$EXPSPACE = AEXPTIME$$
$$EXPTIME = APSPACE$$
$$PSPACE = APTIME$$
$$PTIME = ALOGSPACE$$

Hence, alternation has the effect of shifting the deterministic hierarchy by exactly one level.

4. COUNTING TURING MACHINES

In the preceding sections, non deterministic and alternating TM's have been used only to solve decision problems. However, Valiant [23] and Simon carried out a generalization by considering enumeration problems. Roughly speaking, while decision problems consist in establishing if a structure with a particular property exists in a set of given structures, an enumeration problem consists in counting the number of structures with a given property.

A formal setting for the concept of enumeration - that is fundamental in combinatorial theory - can be established as follows.

Def.1 - An enumeration system is a pair $<I,\mathcal{S}>$ where I is an index set and $\mathcal{S}$ is a system of finite sets indexed by I: $\mathcal{S} = \{S_j | j \in I\}$.

Def.2 - The enumeration function associated with an enumeration system $<I,\mathcal{S}>$ is the function $f: I \longrightarrow N$ defined by $f(j) = \#S_j$.

    (Where, N denotes the set of natural numbers and $\#S$ the cardinality of the set S)

Now, in non deterministic TM's there are, in general different accepting paths for a given input w. Hence, we can associate with every NDTM an enumeration system as follows:

Def.3 - An enumeration system $<I,\mathcal{S}>$ is generated by a non deterministic Turing machine M with input alphabet Σ iff $I = \Sigma^*$ and for every $w \in \Sigma^*$, S_w is the set

    of (different) accepting computations of M on input w.

Def.4 - A Counting Turing Machine (CTM) is a non deterministic Turing machine which (magically) prints the number $\#S_w$ for every input w: so, it computes the enumeration function of the generated system $<I,\mathcal{S}>$.

More formally, we can associate with every node x of an accepting tree T the number N(x) of accepting sequences starting from it as follows:
a)   if x is an accepting leaf, N(x)=1

b)   if x is a non accepting leaf, N(x)=0
c)   if x is internal and has sons $\{x_1,...,x_k\}$, $N(x)=\sum_1^k N(x_i)$.

N(root) is then the number of accepting computation in T.

In [23], Valiant defines the notion of time complexity of a CTM and the class of #P-complete problems. We can extend these definitions to cover also space complexity as follows.

Def.5 - A CTM M has:
- time complexity T(n) iff for every input w of lenght n the <u>longest</u> computation accepting w requires at most T(n) steps;
- space complexity S(n) iff, every computation:
  1) requires at most S(n) cells; and
  2) halts at most after $2^{n \cdot S(n)}$ steps.

Condition (2) is required in order to guarantee the number of computations to be finite; furthermore, it allows to consider all the computations without repetition of configurations.

Def.6 - #P and #PSPACE are the classes of enumeration problems that can be solved by a CTM working in polynomial time or, respectively, space.

Obviously, NP $\subseteq$ #P, since NP contains functions with values in the set $\{0,1\}$ computable by non deterministic TM's.

The notions of polynomial reducibility, and hence of completeness, cannot be directly extended to the classes #P and #PSPACE. In fact, we have to require that the polynomial reduction preserves not only the existence or the absence of accepting paths, but preserves their exact number, i.e. that it be <u>parsimonious</u>.

It is in general easy to prove that the enumeration problem associated with a NP-complete problem is # P-complete. However, in [23] Valiant exhibits a #P-complete problem whose associate decision problem is in P.

Theorem - Let (A_{ij}) be a 0,1 square matrix of dimension n, and let the permanent Per(A_{ij}) be defined by Per$(A_{ij})= \sum_P \prod_i A_{p(i),i}$, where p denotes a permutation of the n-tuple $(1,...,n)$. Then the problem "Calculus of the permanent of A" is a # P-complete problem.

Computing the permanent of A corresponds to counting the number of perfects matchings of the associated graph: the relevant fact is that we can decide on the existence of at least one perfect matching in polynomial time.

In order to exhibit a problem complete in # PSPACE, we can consider the enumeration problem associated with a decision problem which has been shown to be PSPACE-complete,   i.e. the satisfiability problem for quantified Boolean formulas.

Def.7 - A <u>quantified booleans formula</u> (QBF) is a first order boolean sentence φ of the form:
$$\varphi = Q_1 x_1 Q_2 x_2 \cdots Q_m x_m \Psi(x_1 \ldots x_m)$$
where $Q_j \in \{\exists, \forall\}$,   and Ψ is a quantifier-free boolean formula. The variable x_j is said to be <u>universal</u> iff $Q_j = \forall$, <u>existential</u> iff $Q_j = \exists$.

Def.8 - Let $Q = Q_1 \ldots Q_m$ be a sequence of quantifiers $Q_i \in \{\exists, \forall\}$. The set $\mathbb{0}(Q)$ of the <u>assignment trees generated by Q</u> is recursively defined by:
$$\mathbb{0}(\forall) = \{\{0,1\}\}$$

$$\mathbb{D}(\exists) = \{\{0\},\{1\}\}$$
$$\mathbb{D}(\forall Q) = \{0.T_1 \cup 1.T_2 \mid T_1,T_2 \in \mathbb{D}(Q)\}$$
$$\mathbb{D}(\exists Q) = \{0.T \mid T \in \mathbb{D}(Q)\} \cup \{1.T \mid T \in \mathbb{D}(Q)\}$$

where obviously $a.T = \{at \mid t \in T\}$, for $a \in \{0,1\}$

Def.9 - Let $\varphi = Q_1 x_1 \ldots Q_m x_m \ \psi(x_1 \ldots x_m)$ be a QBF; a tree $T \in \mathbb{D}(Q_1 \ldots Q_m)$ is <u>accepting</u> with respect to φ iff for every $t=t_1 \ldots t_m \in T$ we have $\varphi(t_1 \ldots t_m)=1$.

Now, we define the following problem:

SATISFIABILITY OF QUANTIFIED BOOLEAN FORMULAS (#QBF)

INSTANCE: A well formed QBF $\varphi = Q_1 x_1 \ldots Q_n x_n \ \psi(x_1 \ldots x_n)$

QUESTION: Find the number of different assignment trees accepting with respect to φ.

Theorem - #QBF is complete in #P SPACE.

In order to prove this theorem, we show that any problem in # PSPACE can be reduced to an instance of #QBF, by using a construction analogous to the one given by Meyer and Stockmeyer to prove the completeness of the corresponding decision problem in PSPACE. Here, the critical point is that the reduction must be parsimonious, that is it must mantain the number of solutions, while the one shown by Meyer and Stockmeyer is not. The complete proof can be found in [3].

While, as far as complexity classes of decision problems are concerned, it is not known if P-SPACE $\neq$ NP, for enumeration problems the fact that #P-SPACE $\neq$ #P follows from the obvious combinatorial remark that in the class #P-SPACE are contained some problems with 2^{2^n} solutions, n being the size of the input, while for every problem in #P there exists a polynomial P(n) such that the number of solutions is $\leq 2^{P(n)}$. Such a limitation can be overcame if we use ATM's, since the number of subtrees of a tree of depth n is bounded by 2^{2^n}.

Def.10 - A counting ATM (CATM) is an ATM M with an auxiliary device that prints on a special tape the number of accepting subtrees induced by the input.

Def.11 - A CATM has time complexity T(n) if the maximum height of an accepting subtree induced by an input of size n is T(n)

We can associate with every node x of an accepting tree of a CATM the number of the accepting subtrees with root x inductively as follows:
a)   if x is an accepting leaf, N(x)=1;
b)   if x is a non accepting leaf, N(x)=0;
c)   if x is an internal universal node with sons $x_1,\ldots,x_k$, $N(x)= \prod N(x_i)$;
d)   if x is an internal existential node with sons $x_1,\ldots,x_k$, $N(x)= \sum N(x_i)$;

Hence, even for enumeration problems, alternation allows to compute in polynomial time functions that would require polynomial space on non deterministic machines.

5. RANDOM ACCESS MACHINES AND VECTOR MACHINES

On the other hand, the complexity of decision problems has been studied also with respect to models of computation different from Turing Machines, in particular Vector Machines [4] and Random Access Machines [22], which embed implicit parallel features.

Def.1 - A basic random access machine (RAM) consists of:

a) a <u>memory</u> consisting of an enumerable set of registers, with addresses $0,1,\ldots,n,\ldots$, each of which can contain an integer of arbitrary size; the current content of the register n is denoted $\langle n \rangle$; register 0 is used as accumulator.

b) a <u>program</u> P, consisting of a finite sequence of labeled instructions. A table of the admitted instructions, together with their semantics follows; the notation $\langle n \rangle := \langle k \rangle$ is used to mean that the contents of k are copied in n. $\langle\langle n \rangle\rangle$ will indicate the contents of the register whose address is the contents of n: this procedure will be referred to as "indirect addressing".

Instructions	Semantics
<u>a-load</u> n	$\langle 0 \rangle := n$
<u>load</u> n	$\langle 0 \rangle := \langle n \rangle$
<u>i-load</u> n	$\langle 0 \rangle := \langle\langle n \rangle\rangle$
<u>store</u> n	$\langle n \rangle := \langle 0 \langle$
<u>i-store</u> n	$\langle\langle n \rangle\rangle := \langle 0 \rangle$
<u>suc</u>	$\langle 0 \rangle := \langle 0 \rangle + 1$
<u>goto</u> λ	The control is transferred to λ
<u>jzero</u> λ	If $\langle 0 \rangle = 0$, the control is trasferred to λ
<u>halt</u>	The machine stops

Def.2 - A <u>computation</u> of a RAM is as follows:

a) we start with an input $\langle x_1 \ldots x_h, h \rangle$, where $x_1 \ldots x_h \in \{0,1\}^*$ gives the "<u>structure</u>" of the input, and $h = |x_1 \ldots x_h|$ gives its <u>size</u>. The input is stored as follows:

$$\langle 1 \rangle = h$$
$$\langle j \rangle = x_{j-1} \text{ for } 1 \leqslant j \leqslant h+1$$
$$\langle j \rangle = 0 \quad \text{ for } j > h+1$$

b) The program is executed sequentially if no jump <u>(goto)</u> or branching <u>(jzero)</u> instructions are encountered.

c) A special <u>output register</u> z is distinguished. Its content after the machine stopped is the result of the computation.

Def.3 - A RAM M has time complexity T(n) if the longest computation originated by any input of lenght n requires to execute T(n) instructions.

More powerful RAM's can be obtained by adding extra instructions from the following list:

<u>add</u>	n ;	+	$\langle 0 \rangle := \langle 0 \rangle + \langle n \rangle$;
<u>sub</u>	n ;	$\cdot$	$\langle 0 \rangle := \max \{0, \langle 0 \rangle - \langle n \rangle\}$;
<u>mult</u>	n ;	$\overset{.}{*}$	$\langle 0 \rangle := \langle 0 \rangle * \langle n \rangle$;
<u>div</u>	n ;	$\div$	$\langle 0 \rangle := \langle 0 \rangle / \langle n \rangle$; <u>halt</u>, if $\langle n \rangle \geq 0$;
<u>shift</u>	n ;	$\longleftarrow$	$\langle 0 \rangle := \langle 0 \rangle / 2^{\langle n \rangle}$;
<u>and</u>	n ;	$\wedge$	$\langle 0 \rangle := \langle 0 \rangle \wedge \langle n \rangle$, where both operands are considered as binary strings.

In the following, the class of RAM's with the set $\{o_1, \ldots o_n\}$ of additional instructions will be denoted by $RAM(o_1, \ldots o_n)$.

Vector Machines (VM's) [4] are a variant of RAM's in which the registers can contain arbitrarily long sequences of bits (elements of $\{0,1\}$) rather than arbitrarily large numbers. The following operations can be executed on such "bit vectors":

$A \leftarrow K$	the constant bit-vector K is loaded into register A
$A \leftarrow \overline{B}$	"bitwise parallel" boolean complement
$A \leftarrow B \wedge C$	"bitwise parallel" boolean conjunction
$A \leftarrow B \downarrow C$	left shift of B by the distance given by C
$A \leftarrow B \uparrow C$	right shift of B by the distance given by C
$A = 0$	predicate for testing whether A is 0 everywhere.

It is evident that VM's make use of unbounded parallelism, since one computation step (one operation) consists of the execution of an arbitrarily large number of bitwise operations. An analogous remark holds for RAM's, since it is assumed that an operation has unit cost, independently of the value of operands.
These remarks account for the power of RAM's and VM's. In particular, in [4] it is proved that:

$$V_k\text{-PTIME} = \text{PSPACE}$$

where V_k is the subclass of vector machines for which at the kth step in a computation the maximum length of vectors is bounded by 2^{k+n}, where n is the size of the initial vector.

We can now compare classes of RAM's with different sets of operations by using the notion of polynomial reducibility.

Def.4 - A class $\mathcal{R}_1$ of RAM's is polynomially reducible to the class $\mathcal{R}_2$, denoted by $\mathcal{R}_1 \xrightarrow{P} \mathcal{R}_2$, iff for every machine $M_1 \in \mathcal{R}_1$ there exists a $M_2 \in \mathcal{R}_2$ such that M_2 polynomially simulates M_1.

The main results, due to different authors [4,22], are collected in the following diagram:

$$V_k \xleftrightarrow{P} \text{RAM}(+,\dot-,*,\wedge) \xleftrightarrow{P} \text{RAM}(+,*,\wedge)$$
$$\uparrow \qquad\qquad \uparrow P \qquad\qquad\qquad \uparrow P$$
$$\text{RAM}(+,\dot-,*,\dot\div) \xleftrightarrow{P} \text{RAM}(+,\dot-,*,\downarrow) \xleftrightarrow{P} \text{RAM}(+,*,\downarrow)$$

In the next sections we will show that also the classes of machines on the first row can be polynomially reduced to those on the second row, so completing the diagram.

6. RANDOM ACCESS MACHINES AND ENUMERATION PROBLEMS

Arithmetical RAM's are a very natural tool for solving enumeration problems represented by CTM's, since the "auxiliary device" which computes the counting function can be implemented by a program for a RAM. In fact, combinatorial mathematics often expresses the solution of an enumeration problem by means of a "solving formula" based on the usual arithmetic operations. Two examples of such solving formulas are as follows:

Example 1 - $\#\{\text{Permutations of } \{1,2,\ldots,n\}\} = \prod_1^n k = n!$

Example 2 - $\#\{\text{Cycle decompositions of a graph with characteristic matrix A}\} =$
$= \text{per}(A) = \sum A_{1k_1} \cdot A_{2k_2} \cdot \cdots \cdot A_{nk_n}$
where the sum is extended to all the permutations $\{1,2,\ldots,n\}$.

Hence, it is interesting to classify enumeration problems with respect to their complexity in time on RAM's and to compare the classes so obtained with the complexity classes on Turing Machines. In $[3]$, the following fundamental theorem is given:

Theorem – Every enumeration problem in the class $\#$ PSPACE can be solved in polynomial time on a RAM$(+,\dot{-},*,\div)$.

The converse also holds:

Theorem – For every function $f\colon \Sigma^* \longrightarrow N$ computable in polynomial time on a RAM$(+,\dot{-},*,\div)$ there exists a CTM M working in polynomial space whose associated enumeration problem is solved by f.

The proof of this theorem consists in proving that $\#$QBF can be solved in polynomial time on a RAM with the usual arithmetic operations. Being $\#$QBF complete in $\#$PSPACE, this proves the theorem.

Now, remembering that a VM $M \in V_k$ working in polynomial time can be simulated by a RAM$(+,\dot{-},*,\div)$ working in polynomial time, we can conclude that every polynomial time VM can be polynomially simulated by a polynomial time RAM$(+,\dot{-},*,\div)$. A staightforward padding argument allows to generalize the result, so obtaining:

Theorem – $V_k \xleftrightarrow{\;P\;}$ RAM$(+,\dot{-},*,\div)$.

Then, we can complete the previous diagram as follows:

$$V_k \xleftrightarrow{\;P\;} \text{RAM}(+,-,*,\ ) \xleftrightarrow{\;P\;} \text{RAM}(+,*,\downarrow)$$

$$P \updownarrow \qquad\qquad P \updownarrow \qquad\qquad \updownarrow P$$

$$\text{RAM}(+,\dot{-},*,\div) \xleftrightarrow{\;P\;} \text{RAM}(+,\dot{-},*,\ ) \xleftrightarrow{\;P\;} \text{RAM}(+,*,\downarrow)$$

REFERENCES

[1] Aho A.V., Hopcroft J.E., Ullman J.D., The design and analysis of computer algorithms, Addison-Wesley, Reading, Mass., 1974

[2] Batcher K.E., Sorting networks and their applications, Proc. AFIPS Spring Joint Comp. Conf., Vol. 32, 1968, 307-314

[3] Bertoni A., Mauri G., Sabadini N., A characterization of the class of functions computable in polynomial time by Random Access Machines, Proc. 13th ACM STOC, 1981, 168-176

[4] Chandra A.K., Kozen D.C., Stockmeyer L.J., Alternation, J.ACM 28, 1, 1981, 114-133

[5] Cook S.A., The complexity of theorem proving procedures, Proc. 3rd ACM STOC, 1971, 151-158

[6] Dymond P.W., Tompa M., Speedup of deterministic machines by synchronous parallel machines, Proc. 15th ACM Symposium on Theory of Computing, 1983

[7] Floyd R.M., Non-deterministic algorithms, J. ACM 14, 1967, 636-644

[8] Fortune S., Wyllie J., Parallelism in random access machines, Proc. 10th ACM Symposium on Theory of Computing, 1978, 114-118

[9] Garey M.R., Johnson D.S., Computers and intractability, W.H. Freeman and Co., San Francisco, Cal., 1979

[10] Goldschlager L.M., A unified approach to models of synchronous parallel machines, Proc. 10th ACM Symposium on Theory of Computing, 1978, 89-94

[11] Hartmanis J., Simon J., On the power of multiplication in Random Access Machines, 15th IEEE Symp. on Switching and Automata Theory, 1974, 13-23

[12] Hopcroft J.E., Ullman J.D., Formal languages and their relation to automata, Addison-Wesley 1969

[13] Karp R.M., Reducibilities among combinatorial problems, in "Complexity of Computer Computations, Plenum Press, 1972

[14] Karp R.M., Miller R.E., Parallel program schemata, JCSS 3, 1969, 147-195

[15] Kozen D., On parallelism in Turing machines, Proc. 17th IEEE Symposium on Foundations of Computer Science, 1976, 89-97

[16] Muller D.E., Preparata F.P., Bounds to complexities of networks for sorting and for switching, J. ACM, 22, 1975, 195-201

[17] Pratt V.R., Stockmeyer L.J., A characterization of the power of vector machines, JCSS 12 , 1976, 198-221

[18] Rabin M.O., Scott D., Finite automata and their decision problems, IBM J. Res. Dev. 3, 1959, 114-125

[19] Reif J.H., On the power of probabilistic choice in synchronous parallel computations, Proc. 9th ICALP, 1982

[20] Savitch W.J., Relationship between non deterministic and deterministc tape complexities, JCSS 4, 1970, 177-192

[21] Savitch W.J., Parallel and non-deterministic complexity classes, Lecture Notes in Computer Science, 1978, 411-424

[22] Schönage A., On the power of Random Access Machines, Proc. 6th ICALP, Lect. Not. in Comp. Sci. 71, Springer, Berlin, 1979, 520-529

[23] Valiant L.G., The complexity of computing the permanent, Theoretical Computer Science 8, 1979, 189-202

[24] Valiant L.G., Parallel computation, 7th IBM Symp. on Mathematical Foundation of Computer Science, 1982

[25] Yao A.C., On the parallel computation for the Knapsack Problem, Proc. 13-th ACM STOC, 1981, 123-127

ACKNOWLEDGEMENTS

This paper has been supported by Ministero della Pubblica Istruzione, in the frame of the project "Theory of algorithms"

PARALLEL ALGORITHMS - THEORY AND LIMITATIONS

P. Weidner and F. Hoßfeld

Zentralinstitut für Angewandte Mathematik der

Kernforschungsanlage Jülich GmbH

Postfach 1913, 517 Jülich

1. Introduction

Whereas many mental processes are governed by a high degree of parallelism, algorithms were restricted to a sequential flow of operations and data for many decades. This limitation was strongly supported by the serial von Neumann type computer architecture. Favoured by the design and availability of computers with parallel features in several levels and domains, during the last twenty years the design, study and implementation of parallel algorithms gained an increasing interest.

Under rather restrictive models of computation a lot of theoretical results were derived, especially in the field of numerical algorithms. Unfortunately, many of these designs call for an unrealistic high number of processors. Moreover, implementations on existing computers manifest a strong influence of machine organization and data communication on the performance of many algorithms. Therefore, the valuation of algorithms must take these aspects into consideration beside the mere counting of arithmetic operations. The design of specialized array processors is a way to overcome these difficulties through a careful balancing of computation and communication complexity.

2. Theoretical results

After providing some notations and concepts for the representation and valuation of parallel algorithms we will give a short overview of theoretical results, trying to classify problems by the degree of parallelism.

2.1 Notations

To design and study parallel algorithms, one has to fix some model of computation. The following (or some very similar) model can be found in many papers /6, 9, 18/:

  I. The system consists of p individual processors ($p \leqslant \infty$).

  II. The given problems can be represented as sequences of binary arithmetic and logic operations.

III. Each of the p processors can perform anyone of these operations at any time.

IV. Each operation consumes one unit of time.

V. Other processes (data transfer, control etc.) do not consume any time.

VI. There are no data access conflicts.

Whereas this model is rather unrealistic, we adopt it for the outline of some general results. In Section 3 we will discuss these assumptions and show their limits.

To characterize the gain of a parallel computation against the sequential one for the same problem, speedup and efficiency are approved measures. Let $T_p(n)$ be the time complexity of an algorithm for a problem of size n on p processors (n is, as usual in complexity theory, an adequate measure for the input size). Then the speedup and efficiency are defined as $S_p = T_1/T_p$ and $E_p = S_p/p$ respectively with the trivial bounds $1 \leqslant S_p \leqslant p$ and $1/p \leqslant E_p \leqslant 1$.

A graphic representation of algorithms under our model of computation is provided by binary trees. The leaves represent the input values (atoms), the other vertices the operations, the root the final result. Figs. 1 and 2 give an example, the computation of the scalar product $a_1b_1+a_2b_2+a_3b_3+a_4b_4$ on one and four processors, respectively. The speedup is obviously $S_4 = 7/3$, the efficiency is $E_4 = 7/12$.

If the result consists of several values as for the matrix-vector product

$\begin{pmatrix} a & b \\ c & d \end{pmatrix} \cdot \begin{pmatrix} e \\ f \end{pmatrix}$, a corresponding number of trees is needed (Fig.3)

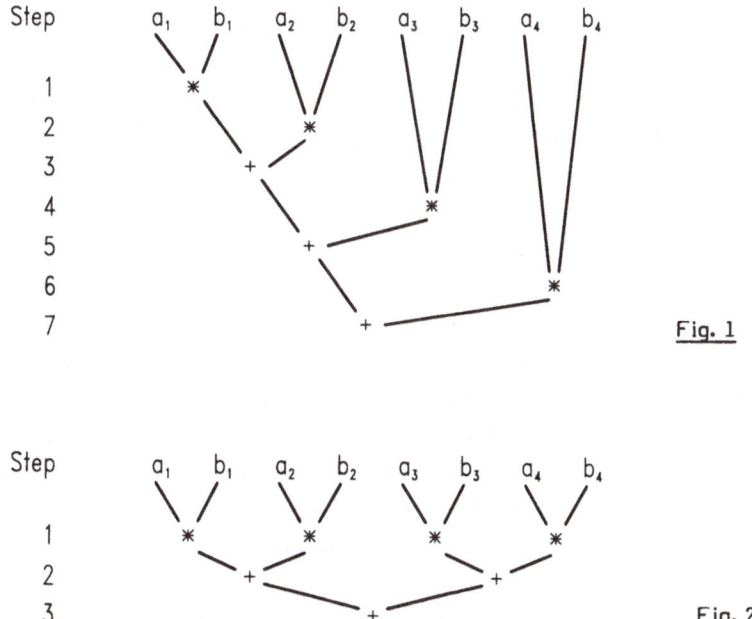

Fig. 1

Fig. 2

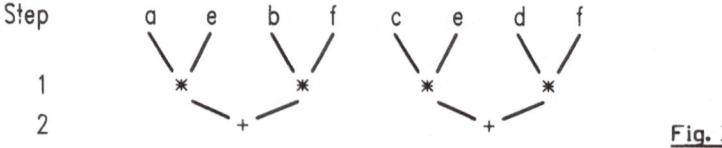

Step

Fig. 3

2.2 Problems with inherent parallelism

Many problems possess a natural parallelism and can be adapted to a parallel
processor model with moderate effort. This is especially true for most problems
of linear algebra /5/. As Figs. 2 and 3 suggest, the scalar product of two
n-vectors and the product of two nxn matrices can be performed in O(log n)
steps using n or n^3 processors, respectively (all logarithms used in this paper
are to the base 2). Similarly, the Gaussian algorithm (without pivoting) for
the solution of n simultaneous equations in n unknowns needs O(n) steps on
$O(n^2)$ processors. Using the Cayley-Hamilton identity, Czanky /5/ proposed a
different algorithm for the solution of systems of equations and matrix
inversion which needs only $O(\log^2 n)$ steps on $O(n^4)$ processors. Beside the
excessive number of processors, the method turns out to be numerically
unstable. This example shows that mere counting of operations is not always
sufficient for the valuation of an algorithm, further criteria have to be
applied.

Numerical methods for the solution of linear elliptic differential equations
represent another class of algorithms well suited for parallel processing.
Discretization of the differential equations by finite differences or finite
elements leads to large systems of simultaneous linear equations which can be
handled by the above mentioned methods. Besides, due to the sparseness of the
systems, iterative methods can be applied. Here, it turns out that Jacobi-type
methods are easier to parallelize than Gauss-Seidel and related techniques
/16/; in this area numerical research is strongly pushed by the advent of
parallel processors. The same is true for solution methods for parabolic
equations /15/.

2.3 Problems not obviously parallelizable

Not all problems possess the extensive independence of input data and
operations, which makes linear algebra methods, for instance, so well suited
for parallel processing. Recurrence relations are an extreme example for
computations with strong data dependencies: an actual value is computed from
its direct predecessors. On the other side, many important problems, such as
polynomial evaluation or the solution of triangular systems of equations, can
be done very efficiently by recurrence methods, at least on a single processor.
Therefore, a lot of research was realized to construct efficient parallel

algorithms for recurrence calculations /9/.

We illustrate one technique for linear recurrences $R(n,m)$ which can be written
as
$$x_i = 0 \qquad \text{for } i \leq 0$$

$$x_i = c_i + \sum_{j=n-m}^{i-1} a_{ij}x_j \qquad \text{for } 1 \leq i \leq n$$

with given values c_i, a_{ij}.

$n-1$ processors are associated with $x_2,\ldots,x_n$ and initialized with $c_2,\ldots c_n$ respectively. In the first step the given value $x_1 = c_1$ is broadcasted to all processors, multiplied by a_{1i} and added to c_i respectively. In processor number two now x_2 is available and can be treated as x_1 before. This procedure uses $T_p = 2(n-1)$ steps of multiplication and addition on $p=n-1$ processors, compared to $T_1 = m(2n-m-1)$ operations in the sequential case. The resulting speedup

$$S_p = \frac{m}{2}\left(1 + \frac{n-m}{n-1}\right)$$

is equal to $\frac{n}{2}$ for $m=n-1$, which is rather satisfactory; this case applies to triangular systems of equations. For $m=1$, e.g. Horner's rule for polynomial evaluation, $S_p = 1$ means no gain at all. For small m other more effective methods, like the product form algorithm /9/ exist.

Some tasks seam not to fit into our model of computation, at least at first sight, but can be transformed to other problem classes. Graph problems are here a striking example. Graphs can be represented in various ways by matrices, e.g. the adjacency matrix, and many problems can be reduced to algebraic operations with these matrices. In /14/ the determination of the length of shortest paths between all vertices of a weighted directed graph is treated as an example. For sequential processing, these methods are inferior to the usual graph algorithms, they are superior for parallel processing, however.

A large class of methods in numerical mathematics is characterized by the fact that the principal amount of computational work is consumed by the evaluation of more or less complicated given functions. Optimization, quadrature, solution of ordinary differential equations and some others fall into this category. Simpson's integration formula, for example, consists of evaluations of the integrand at several points and some multiplications and additions of these function values. In any case, the function evaluations are the most time consuming part. It is not possible to give general rules for the parallelization of these methods. Depending on the type of the given function a simultaneous evaluation at different points can yield a reasonable speedup, but sometimes also a parallel method for a single evaluation can be promising.

At present, there are some problem classes where parallel algorithms are not yet developed because of inherent sequential behaviour or very strong dependencies between the data. Adaptive methods of numerical mathematics must be mentioned here. The succession of operations depends on the course of intermediate computational results and cannot be fixed in advance.

The now very popular Monte Carlo methods for the simulation of solid state behaviour, for instance, seem also to be difficult to parallelize, although some first steps to progress were published /13/.

3.   Problems of organization and data communication

The results of the second section were derived under the assumption of an idealistic model of computation and did not pay attention to possible implementations on existing machines. Due to the variety of built and designed computers with more or less parallelism it is impossible to set up universal design principles for parallel algorithms. Nevertheless, there are some difficulties and as well remedies to overcome them, which apply to larger classes of computer organizations. We try to outline some of these aspects.

3.1 Influence of SIMD/MIMD-organization on algorithms

The design principles for parallel algorithms on SIMD (Single-Instruction-Multiple-Data stream) computers - we incorporate also pipelined execution into this category - and MIMD (Multiple-Instruction-Multiple-Data stream) computers have to follow rather different lines. On SIMD machines (e.g. ILLIAC IV, CRAY-1S, ICL DAP) it is desirable that large numbers of independent data are processed by the same sort of operation. On a processor array like ILLIAC or DAP this is caused by the fact that all processors perform the same operation in parallel; the data have to be independent to minimize communication between the processors. On pipeline machines like CRAY-1S good performance is achieved if the time for setting up and filling the pipeline can be neglected, which requires a pretty long sequence of equal operations; here, independence of data is necessary since operands are "hidden" for several time steps in the pipe.

Linear algebra computations with large vectors and matrices are generally well suited for this organization.

On the contrary, algorithms for MIMD-organized processors (e.g. HEP, Cm*) should consist of larger disjoint portions that can be treated independently and need only occasional synchronization and data communications. Some simulation methods /2/ and the chaotic relaxation method for the solution of partial differential equations /1/ are typical examples.

3.2 Communication complexity

The assumption V. of section 2 either restricts severely the types of problems which can he parallelized or leads to an unrealistic performance behaviour of algorithms. This is best illustrated by a result due to Gentleman /4/ about the communication complexity of matrix multiplication. Two nxn matrices are to be multiplied. The model of computation is a rectangular array of processors with nearest neighbour interconnections and private memory. At the beginning of the computation, each element of the given matrices is represented once and only once within the machine and no two elements of the same matrix are stored in the same private memory. These are rather realistic conditions if not an abundance of memory space is available. Then it turns out that for large n at least $0.35n$ data movement steps are necessary to perform the multiplication, independent of the chosen algorithm and the number of processors. For large n this effort exceeds considerably the achievable minimum number of arithmetic steps, namely $1+\log n$. This example shows that even for problems in linear algebra the communication complexity has to be taken into account beside the computational complexity.

3.3 Access conflicts

If several processors share a common memory, access conflicts occur since the bandwidth of data paths is bounded. To reduce these conflicts adapted memory organizations have to be created. Again, we take matrix operations as an example to illustrate the method. If a 4x4 matrix is stored in the usual way in four parallel memory units, simultaneous access to rows or diagonals, but not to columns, is possible (Fig. 4).

memory unit

1	2	3	4
a_{11}	a_{12}	a_{13}	a_{14}
a_{21}	a_{22}	a_{23}	a_{24}
a_{31}	a_{32}	a_{33}	a_{34}
a_{41}	a_{42}	a_{43}	a_{44}

Fig. 4

Storing the data in 5 units as indicated in Fig. 5 allows parallel access to rows, columns or antidiagonals. This 'skewing' procedure was systematically investigated in /12/. More sophisticated schemes - with certainly more redundancy - permit parallel access even to diagonals and square blocks.

Generally, the selected scheme depends on the sort of calculations to be done and is a compromise between available storage and acceptable access conflicts.

memory unit

1	2	3	4	5
a_{11}	a_{13}	–	a_{12}	a_{14}
a_{22}	a_{24}	a_{21}	a_{23}	–
a_{33}	–	a_{32}	a_{34}	a_{31}
a_{44}	a_{41}	a_{43}	–	a_{42}

Fig. 5

3.4 Effects of pipelining

Pipeline architectures offer only a very limited degree of parallelism: the maximum possible speedup is given by the number of stages in the pipeline. Nevertheless, many of the design principles for parallel algorithms yield efficient algorithms for pipeline computers.

Whereas a pipelined processor performs the operations sequentially, there is a delay between data input and availability of the result, proportional to the number of pipeline stages. This fact is especially unfavourable for the computation of recurrences. However, in some cases this difficulty can be overcome by suitable transformations of the recurrence. We illustrate this technique on a simple example, a Fibonacci-type recurrence, which has the usual form

$$
\begin{aligned}
x_n &= \ldots \\
x_{n+1} &= \ldots \\
x_{n+2} &= x_{n+1} + x_n \\
x_{n+3} &= x_{n+2} + x_{n+1}
\end{aligned}
$$

For the determination of x_{n+2} the values calculated immediately before x_{n+1} and x_n are needed.

Through simple insertion the recurrence can be transformed to the equivalent form

$$
\begin{aligned}
x_n &= \ldots \\
x_{n+1} &= \ldots \\
x_{n+2} &= \ldots \\
x_{n+3} &= 3x_n + 2x_{n-1} \\
x_{n+4} &= 3x_{n+1} + 2x_n
\end{aligned}
$$

which is now suitable for a 3-stage pipeline, since only the values computed three steps before are needed. This method and similar ones can be extended to more general recurrences with constant coefficients.

4. Algorithmic hardware

The aspects and examples of section 3 suggest that for the design of parallel algorithms a unity of the complexity of computation, communication and control is desirable. Here, universality of the computer model and largeness of the problem class are in unsurmountable conflict. A possible alternative is the design of so called algorithmic hardware, which means the construction of architectures for special tasks or problem classes where arithmetic steps, communication and control functions are carefully harmonized. We illustrate this idea with three examples.

4.1 Systolic arrays

Systolic arrays were proposed by Kung and Leiserson /11/ as a special purpose hardware, realizable in VLSI technology, to perform matrix computations. There are several types of arrays with different interconnection structure, dependent on the task to be handled. We describe here a simple case, namely a linear array to compute the matrix-vector product for a band structure matrix.

In general, the product $y = Ax$ can be written as a recurrence

$$y_i^{(1)} = 0$$
$$y_i^{(k+1)} = y_i^{(k)} + a_{ik}x_k \qquad\qquad 1 \leq i \leq n$$
$$y_i = y_i^{(n+1)}$$

If the matrix A has bandwidth w, a linearly connected array of w processors performs the computation in the following way (Fig. 6).

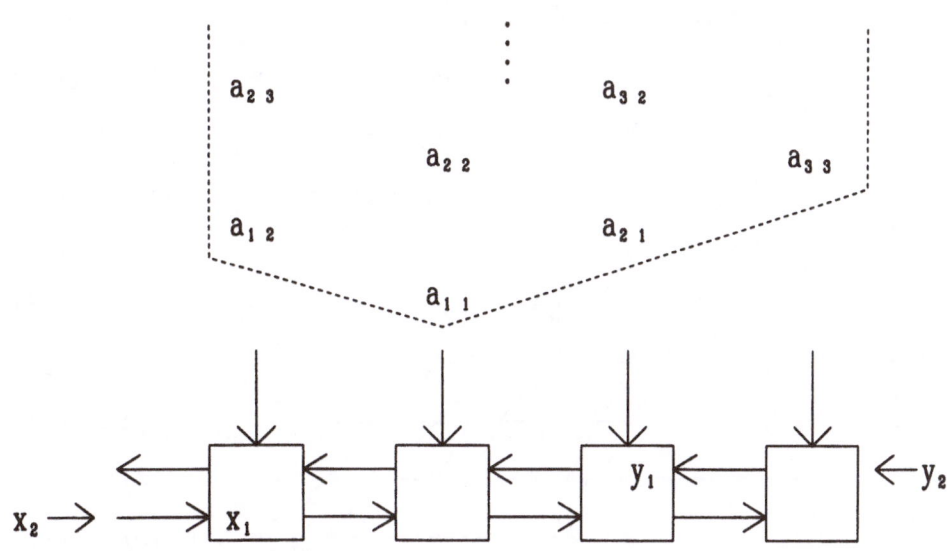

Fig. 6

All entries are moving in a pipelined manner: the y_i, initially zero, to the left, the x_i to the right and the a_{ij} down, all moves being synchronized. Each y_i accumulates all its terms following the above recurrence and has the correct value if it leaves the array. A more detailed analysis of the algorithm shows that every two steps each processor has to perform simultaneously three shift operations and one multiply-add. It follows that the whole product can be computed in T_p = 2n+w time units on p=w processors, compared to T_1 = 0(wn) in the sequential case. Here, communication (shift operations) and computation (multiply-add) run with the same "pulse".

4.2 FFT-Processors

The Fourier transform plays an important role not only in signal processing and similar fields, but also for the solution of partial differential equations.

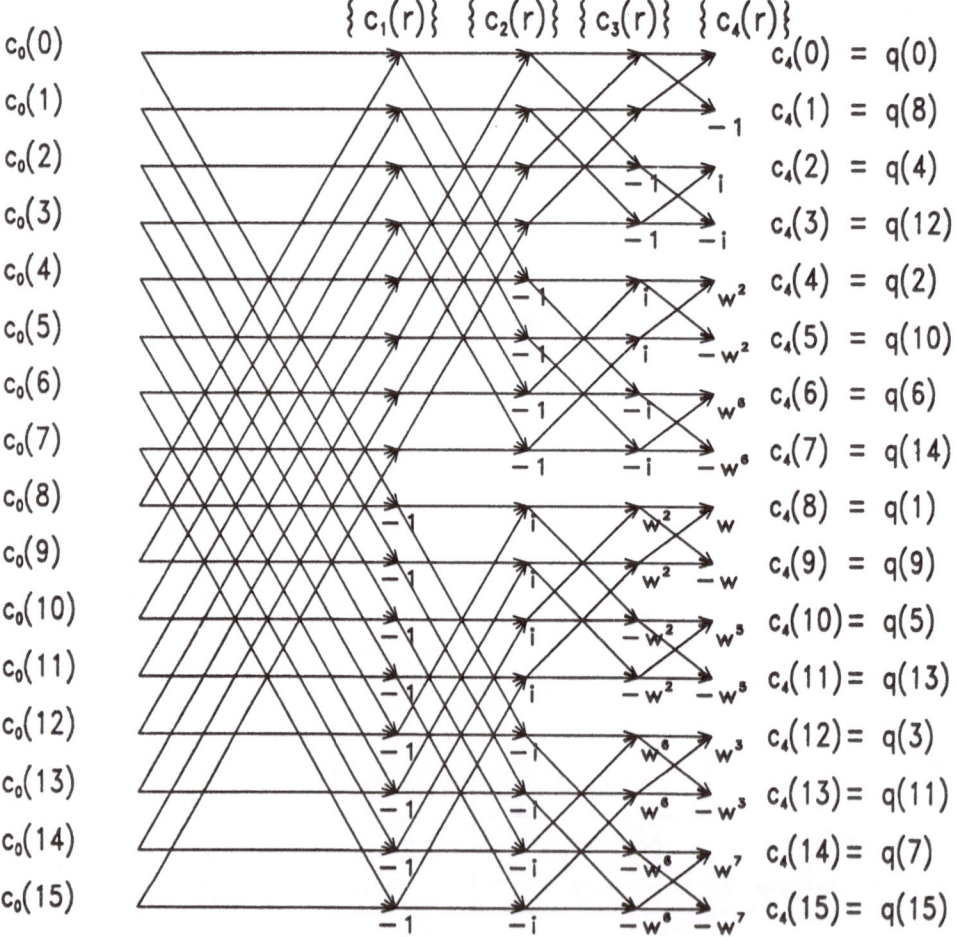

Fig. 7

The discovery of the fast Fourier transform (FFT) represents one of the major breakthroughs in the design of efficient algorithms /3/. Already in the original paper, the inherent parallelism of the method was pointed out. There exist now several implementations of the original and derived algorithms on SIMD architectures /15/. Also systolic arrays are well suited for the FFT. The flow graph for the transform (Fig. 7) suggests the use of a perfect shuffle network with suitably placed processors /17/. This is realized in Fig. 8. In this way, data movement and arithmetic steps are perfectly consistent.

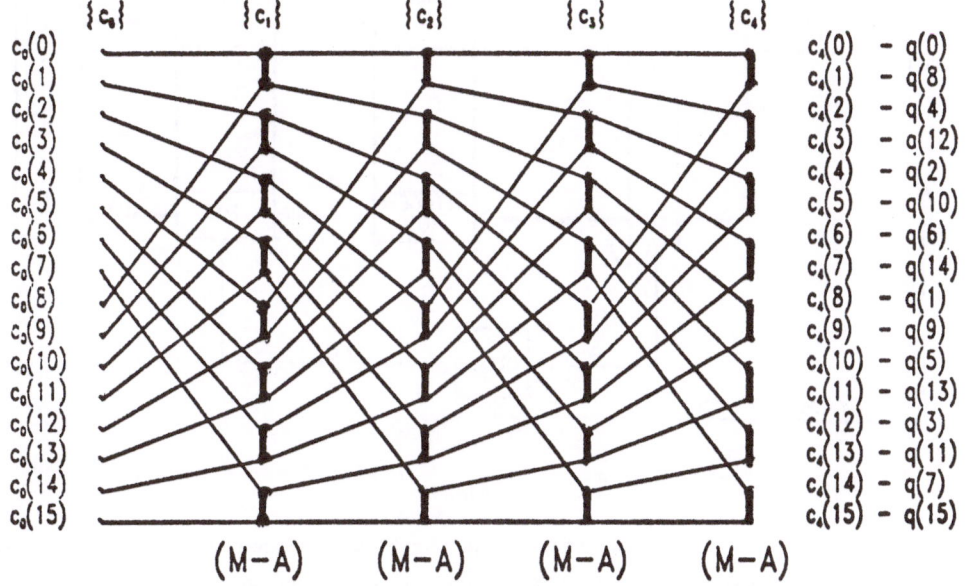

Fig. 8

4.3 Evaluation of CTOF-experiments

Correlation time-of-flight (CTOF) methods are a promising technique for beam spectroscopy. In the simplest case, a stationary monochromatic primary beam of particles is periodically modulated according to a pseudo random binary sequence (a_i) with period N before it is scattered by a target whose characteristics are to be investigated. The output signal Z_i (at discrete time) and the scattering spectrum S_r describing the properties of the target are related by

$$S_r = k_1 \sum_{i=1}^{N-1} z_i a_{i-r} + k_2 \sum_{i=0}^{N-1} z_i + k_3 \quad , \quad 0 \leqslant r \leqslant N-1$$

with suitable constants k_1, k_2, k_3.

In /7/ we proposed a parallel hardware realization for the online evaluation of this formula (Fig. 9).

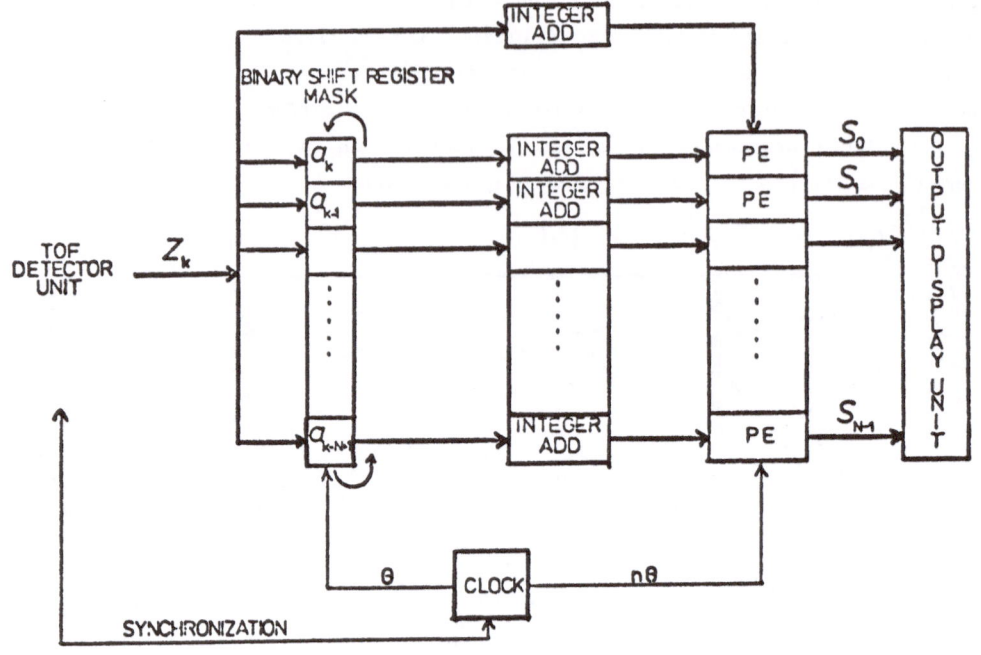

Fig. 9

The counting rate Z_i is broadcasted to N positions, masked with a binary shift register mask according to the sequence (a_i) and added up in parallel. The sum Z_i is computed in a separate integer add unit. Only after N such steps the processing elements (PE) are activated to figure out the whole expressions for S_r ($0 \leq r \leq N-1$). The control consists of the activation of the shift and integer adds in every step and the activation of the PE's after N steps. This control can follow fixed time stamps or can be event driven. In /7/ architectures for further CTOF methods are described.

The shift register technique can be used for the parallel realization of discrete convolution transforms, which are important for signal processing.

5.   Conclusions

Whereas a relatively young discipline, parallel algorithms were extensively studied and a lot of deep results were established. Nevertheless, a general framework for the design and representation of parallel algorithms is still missing.

Another open problem is the adaptation of algorithms designed under the assumption of unbounded parallelism to real architectures consisting of a fixed number of processors. The theorem of Brent /6/ and its proof gives some advice how to manage this, but the resulting algorithms are only crude simulations on a finite number of processors and therefore not very efficient. There exist some promising concepts for matrix problems /8/.

As the example of Czanky's method for matrix inversion shows, for numerical problems the question of stability has to be investigated for parallel algorithms. The few examples treated in the literature indicate that parallelization can create or destroy stability /10/.

In the domain of communication complexity and interconnection networks the interference and compatibility with computational complexity is investigated and understood only in some special cases.

This enumeration of research topics is far from being complete and is only intended to give a taste of the liveliness of the field and the urgent need for future investigations.

References

/1/ Baudet, G.M.: Asynchronous iterative methods for multiprocessors, J. ACM 25 (1978), 226-244

/2/ Bucher, I.Y., Buzbee B.L. and Frederickson, P.O.: Experiment in parallel processing a large scientific code, Proc. 1981 Int. Conf. on Parallel Processing, 1981, 166-167

/3/ Cooley, J.W. and Tukey, J.W.: An algorithm for the machine calculation of complex Fourier series, Math. Comput. 19 (1965), 297-301

/4/ Gentleman, W.M.: Some complexity results for matrix computations on parallel processors, J. ACM 25 (1978), 112-115

/5/ Heller, D.: A survey of parallel algorithms in numerical linear algebra, SIAM Review 20 (1978), 740-777

/6/ Hoßfeld, F.: Parallele Algorithmen, Reihe Informatik-Fachberichte, Bd. 64. Berlin-Heidelberg-New York: Springer 1983

/7/ Hoßfeld, F. and Weidner, P.: Parallel evaluation of correlation time-of-flight experiments, CONPAR Nürnberg, 1981, W. Händler ed., 441-452

/8/ Hwang, K. and Cheng, Y.-H.: Partitioned matrix algorithms for VLSI arithmetic systems, IEEE Trans. Computers C-31 (1982), 1215-1224

/9/ Kuck, D.J.: The Structure of Computers and Computations, Vol. 1. New York: J. Wiley & Sons 1978

/10/ Kuck, D.J., Lawrie, D.H., Sameh, A.H. (eds.): High Speed Computer and Algorithm Organisation. New York: Academic Press 1977

/11/ Kung, H.T., Leiserson, C.E.: Systolic arrays (for VLSI), Proc. SIAM Sparse Matrix 1979, 256

/12/ Lawrie, D.H.: Access and alignment of data in an array processor, IEEE Trans. Computers C-24 (1975), 1145-1155

/13/ Oed, W.: Monte Carlo simulations on vector machines, Angewandte Informatik 24 (1982), 358-364

/14/ Reghbati (Arjomandi), E., and Corneil, D.G.: Parallel computations in graph theory, SIAM J. Computing 7 (1978), 230-237

/15/ Rodrigue, G. (ed.): Parallel Computations. New York: Academic Press 1982

/16/ Schultz, M. (ed.): Elliptic Problem Solvers. New York: Academic Press 1981

/17/ Stone, H.S.: Parallel processing with the perfect shuffle, IEEE Trans. Computers C-20 (1971), 153-161

/18/ Traub, J.F. (ed.): Complexity of Sequential and Parallel Numerical Algorithms, New York: Academic Press 1973

INTERCONNECTION NETWORKS FOR MIMD MACHINES

L. Ciminiera   C. Demartini   A. Serra

CENS -   Diparimento di Automatica e Informatica - Politecnico
di Torino - Corso Duca degli Abruzzi, 24 10129 Torino

1. Introduction

One of the most fruitful techniques used to increase the processing speed is
based on the concurrent or parallel execution of different tasks. It can be
verified that in many real-time applications such as image processing, and weather
computations, where an instruction execution rate of more than one billion floa-
ting- point instructions per second are required, concurrent processing is
necessary. With the advent of VLSI technology, it has become economically feasible
to construct processing system by interconnecting hundreds even thousands of
processors and memory modules. The user writes application programs using a
parallel language: processes are generated by compiling and partitioning those
programs. Each process is assigned to an individual processor; since the processes
should cooperate, the corresponding processors need to exchange block of data
through some interconnection facility. The communication subsystem must be
carefully designed since in concurrent systems, a large amount of computing power
avails could not be exploited, if each processor must spend a lot of time for
sending or receiving data to or from other processors.
From the implementation point of view, many parameters must be evaluated in order
to choose a cost-effective communication network. They concern network topology,
control strategy, switching method and operation mode.

2. Network Topologies

Network topology is an important parameter for developing a suitable
architectural structure and many topologies have been considered for telephone
switching connections. Fig. 1 describes the topologies of interconnection
networks. Static topologies can be classified according to dimensions required for
layout: one-dimensional two dimensional, three dimensional and hypercube as show
in Fig. 1. An example of one-dimensional topology is the linear array used for
some pipeline architectures. Two dimensional topologies include the ring, star,
tree /1/, near neighbor mesh /2/ and systolic array /3/. Three dimensional
topologies include the completely connected, chordal ring and 3-cube connected
cycle /4/ networks. A D-dimensional W-wide hypercube contains W nodes in each
dimension and there is a connection to a node in each dimension. The cube
connected-cycle is a derivation of the hypercube. The graph of the 3-cube-connec-
ted cycle can be obtained by replacing each node of the 3 cube by a 3-node cycle.
Each node in the cycle is connected to the corresponding node in an adjacent
cycle. Topologies in the dynamic subclass can be divided into three classes:
single stage, multistage and crossbar.
A single stage network is obtained if a stage of switching elements is used along
with a link connection pattern. The shuffle exchange network /5/ is a single stage
network based on a perfect shuffle connection cascaded to a stage of switching
elements. Since each input can reach in a sigle pass through the network only a
limited number of outputs, multiple passes are often required for transmitting
data.
A multistage network is composed by more than one stage of switching elements, and
they can usually connect an arbitrary input-output pair in a single pass. They can
be divided in one-sided and two-sided. The one-sided network have the output ports

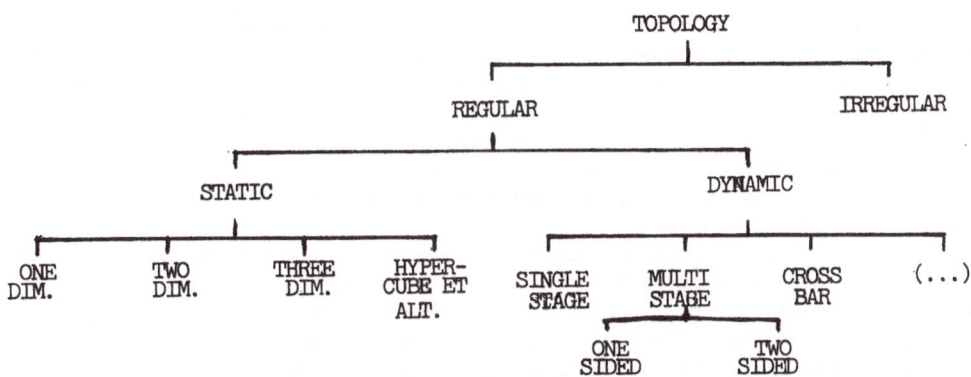

Fig. 1. Classification of the interconnection topologies.

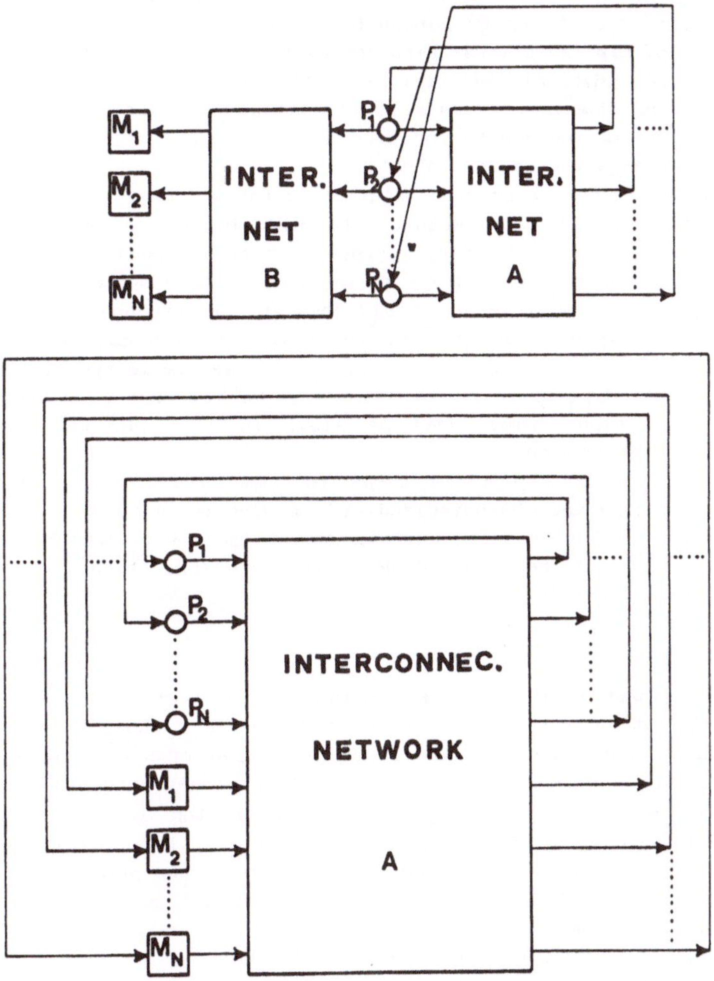

Fig.2. Two possible implementations of a multiprocessor system using interconnection networks.

connected to the same set of devices, which are connected to the input ports.
The two-sided multistage networks have different devices connected at the input
and output sides. They can be divided into three classes: blocking, rearrangeable,
non blocking. In blocking networks, when two or more input devices try to
establish a connection, conflict may arise in the user of network resources.
A rearrangeable network can perform any possible connection between input and
output by rearranging its existing connection, so that path for a new input output
pair can always be established.
Strictly non blocking networks always allow a new request to establish the
connection without rearranging the connections in progress.
Fig. 2 shows how it is possible to use one-sided and two-sided network for
building a multiprocessor system. In general, two networks are required; one for
processor to processor and one for processor to memory communications. In Fig. 2a,
the former is implemented by means of an one -sided network, while the latter is a
two-sided network. The two network may also be merged in a single one-side network
as shown in Fig. 2b. Hovever, this second solution is very expensive since the
cost of a network is O (N log N), where N is the number of inputs. On the other
hand, the scheme of Fig. 2b allows also memory to memory transfers.
The least expensive physical form available for the processor- memory switch is
the time-shared bus. However, it is well known that the time-shared bus has a low
transfer rate which is inadequate for even a small number of processors, if they
have to exchange data with a moderate rate.
On the other hand the most expensive solution is the adoption of the full crossbar
switch, which can also be considered as the most complex switch (O (N^2)). If we
consider the current low cost of microprocessors and memories, a crossbar would
probably cost more than the rest of the system components combined.
The absence of a switch with a reasonable cost and performance has prevented the
growth of large multiprocessor systems. To avoid the high cost of switch, some
loosely systems can be defined where sharing of main memory is restricted.
The processor can access directly and fast only private memory while many other
reference to the common memory may be slow, indirect and may even involve
operating system intervention.
The solution to all these problems seems to come from a class of connecting
schemes /6/ /7/ which are characterized by a medium complexity $O(N \log_2 N)$ and
performances close to the crossbar switch. One of the most interesting classes of
networks of this type is the class of delta networks /6/, which includes most of
the networks previously defined.

2.1 Delta Networks

In order to understand the basic principle involved in the development of
delta networks, let us consider a 2x2 crossbar switch, shown in Fig.3. It can
connect the upper input to either the lower, the upper or both outputs, according
to the value of some control bits. In delta networks only conditions a and b are
considered. If the control bit is O then the switch is set with straight
connections and if 1 it is set with cross connection. Hence each input can be
connected to either output. Following the previous description and considering
only one input we can build a 1-by-2^n demultiplexer using the 2x2 crossbar
switches connected in the form of a binary tree. In Fig. 4a a 1-by-8 demultiplexer
tree is reported in which destinations are marked in binary. If the source
requires to connect to destination $(d_2 d_1 d_0)_2$ then the root node is controlled by
bit d_2, the second stage modules are controlled by bit d_1 and the last stage
modules are controlled by bit d_0.

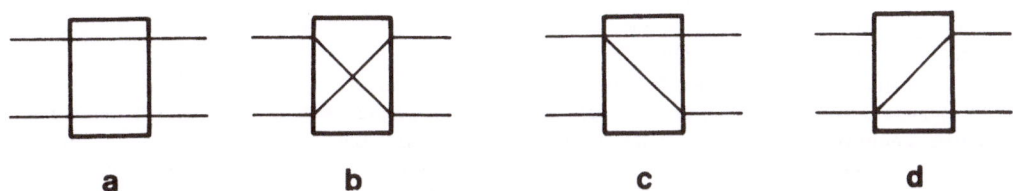

Fig.3. The four possible states of a 2x2 crossbar switch.

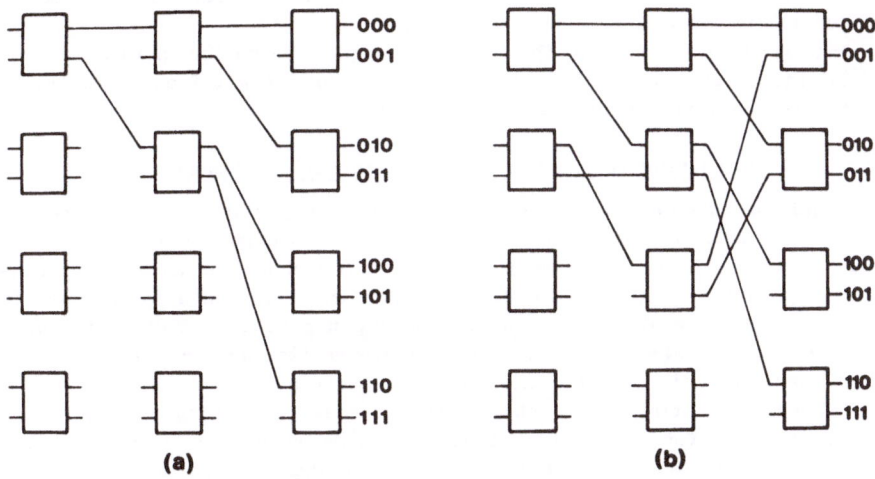

Fig.4. First two steps in the construction of delta network.

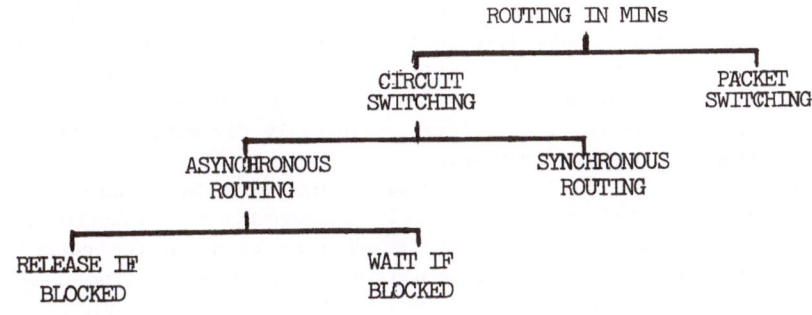

Fig.5. Alternatives for implementing routing in MINs.

We can see that the input chosen can be connected to any the eight output terminals and, obviously, the lower terminal input of the roor node can be connected to any of the output in the same manner.

Now we consider another aspect of the basic 2x2 module: it is assumed that it has the capability to arbitrate between conflicting requests. In other words when both inputs require the same output terminal then only one of them will be connected and the other will be blocked or rejected. Indeed, given the building mechanism based on a tree, it turns out that a single path exist between each input-output pair, ando no detour is possible. If we want to construct an 8x8 network using 2x2 switches we employ the principle used previously to obtain a 1-by-2^n demulti-plexer. Every additional input must also have its own demultiplexer tree to connect to any of the eight outputs. So we start with a demultiplexer tree, then for each additional input we build another demultiplexer tree on the partially constructed network. For doing that we can use the already existing liks as part of the new tree and add extra links and modules if needed. Such procedure is continued until the network is completed.

The network we have developed is called a digit controlled network or delta network. In fact each module is controlled by a single digit from the destination address, no other external controls are required.

3. Routing techniques in digit controlled networks

Routing algorithms can be divided into two classes: algorithms which implement the circuit switched mode and algorithms which implement the packet switched mode. In the circuit switched mode, once the path is established the modules or interchange boxes in the path remain in their specified state until the path is relased. Thus, there is a complete circuit established, from input port to output port, for that path. In the packed switched mode, a packet makes its way from stage to stage releasing links and interchange boxes immediatly after using them. Thus only one module is used at a time for each message. Circuit switching must be used in networks constructed for combinational logic, where there are no buffers in the interchange boxes for showing delayed data. Fig.5 depicts the classification of routing tecniques in multistage interconnection networks.

3.1 Packet switching technique.

Let us consider an implementation of the packet switching node for the cube-network.

Packet switched mode is possible if there are buffers or queues in the interchange boxes. We assume the network is able to operate in on SIMD or MIMD environment.

According to the last assumption it is necessary to provide for buffering of data in each module or interchange box. In this way the trhroughput of the network is improved and the number of conflicts considerably diminuished, particularly if we are operating in the MIMD environment.

The advent of LSI and VLSI chips allows to implement one module in a single VLSI chip. In this way the module can be made considerably more intelligent then simple combinational logic for example it is possible to queue for buffering packets. Fig.6 reports the block scheme of a possible VLSI implementation of an interchange block in the cube-network /12/. Two input queues are provided for buffering incoming data. The control unit has the task of arbitrating conflicts in the switch setting requests, handshaking with the other interchange boxes (to which it is connected), and controlling the tag modification units.

Even if the input queues are projected as linear queues and could be implemented as first-in-first-out (FIFO) buffers, the delay incurred by passing through each

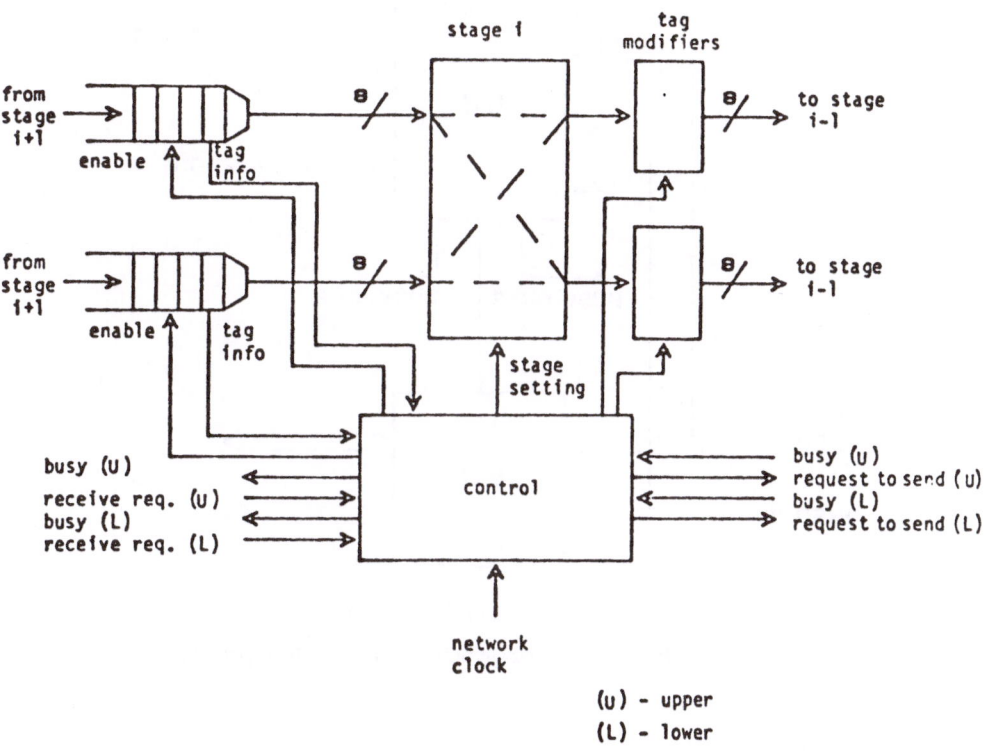

Fig. 6. Diagram of a 2x2 crossbar switch implementation for
         for packet switching.

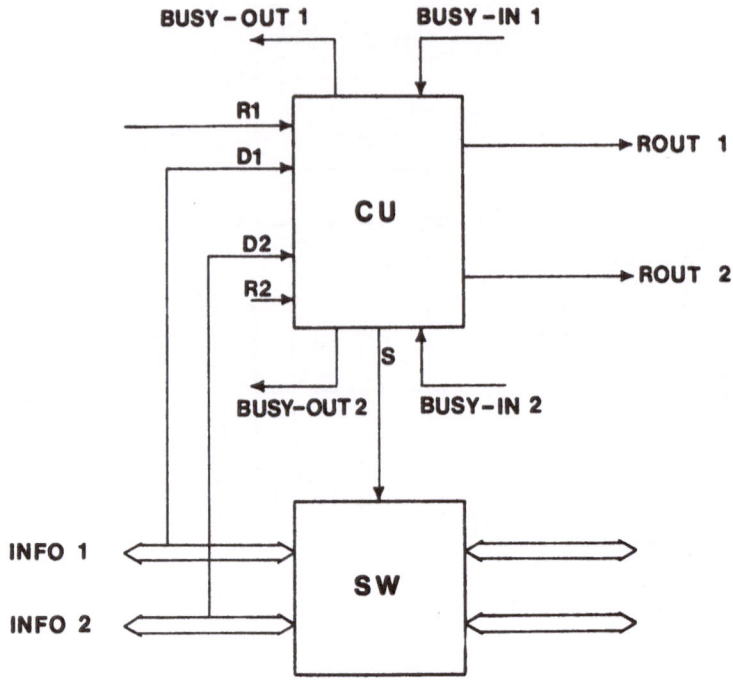

Fig. 7.   Block Diagram for synchronous and RAR protocols.

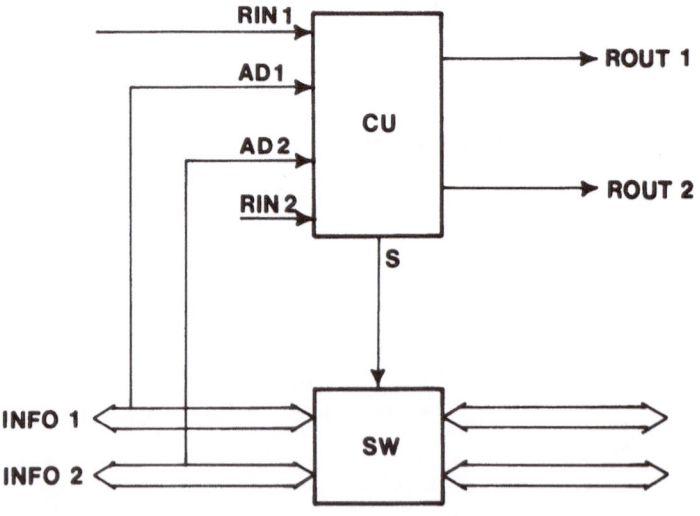

Fig. 8.   Block diagram for WT protocol.

location would be undesirable. Such queues can be implemented using pointers into two circular buffers, in this way it is possible to simulate the functions of the queues. Four pointers should be required, one to point to the front and one to point to the back of each queue.

There are two input and two output data paths. Two control lines are associated with each data path for the busy and receive signals.

3.2 Synchronous circuit switching technique.

Let us consider the implementation of delta networks /6/. Fig.7 reports the functional block diagram of a 2x2 crossar module of a delta network. All single lines in the figure are one bit lines. The double lines on SW box, represent address lines, incoming and outcoming data lines, read/write control line. The function of the SW box is that of a single 2x2 cossbar: if the input s is 1 then a cross connection is obtained, on the other if s is 0 then a straight connection is configured. The function on the CU box is to generate the signal S and provide arbitration. A request is generated at an input port if the corresponding request line is 1. The destination digit provides the nature of the request: when it is 0 the connection to upper output part is obtained, when it is 1 the connection to the lower part is generated. If a conflict arise, the request RIN1 is given the priority and a busy signal can be transmitted to the source which originated the blocked request. The logic equations for all the labeled signals are as follows:

S=RIN1 AD1+RIN2 AD2 RIN1; ROUT1=RIN1 AD1+RIN2 AD2;
ROUT = RIN1 AD1+RIN2 AD2; BUSYOUT1=S BUSYIN2+S BUSYIN1; BUSYOUT2=S BUSYOUT1+S BUSYOUT2

We now examine the operations on a $2^n x2^n$ delta network usign the above 2x2 modules.

All processors which require memory access can place eventually at the same time a 1 on the respective request lines. After a delay equal nt where t is the delay due to the single module, the busy signals are valid. If the busy line is 1, then the processor must try again to submit its request, that means that the processor continue to hold the request line high.

The processor which obtained the communication path (busy signal is 0) can have valid read-data after a delay equal to t*n plus the memory access time. The implementation described here is synchronous in the sense that all the requests are issued at fixed intervals and at the same time. In other words we can say that the time is devided into slots with equal duration and, at the beginning of each slot, all the pending requests are submitted to the network. The generated requests which cannot find a free path to their destination, must submit the request at the next time slot.

One of the main drawbacks of synchronous routing technique is that each connection is mantained only for one time-slot; if the amount of data to be transmitted by the requesting processor is very long, the processor itself must break down the message into packets which can be transmitted into a single slot. This requirement increases the complexity of the transmission protocol to be supported for communication management.

However there is a considerable advantage from the hardware point of view; only combinational circuits are needed since the input requests do not change within a time slot.

3.3 Asynchronous circuit-switching techniques

Since delta networks fall into the class of the blocking networks, during the pathfinding process, a request for a connection may try to obtain a trunk occupied by a different connection. In this case the new request cannot be honored. The blocked request can be processed using different strategies; the following sub-sections deal with the implementation and the comparison of two of such strategies.

3.3.1 Waiting policy

The most straightforward policy for dealing with blocked request is the waiting (WT) policy. In this case, the blocked request waits until the requested trunk is available for continuing the pathfinding process. Of course, the previously-occupied trunks, constituting the path from the transmitting processor up to the crossbar where the request is blocked, are held (channel holding) since, in circuit-switched networks, the trunks occupied can be released only if the data transfer is terminated. Since channel holding phenomena occur when the WT policy is used, a request A, blocked at stage i ($o < i \leq n$), may wait for a trunk held by another request, B, blocked at stage j ($j > i$). Therefore, the request A can set up the connection only if the data exchange associated with the request B is terminated. However, it may happen that, eliminating only the request B, the request A can set up the complete connection; in this case, A is blocked by a blocked request.

Since a request may be blocked by another blocked request, it is worth-while proving that no deadlock situations occur. This can be done by showing that there is a maximum waiting time before the whole connection is set up.

__Theorem__. Given a delta network having N inputs and outputs, the maximum value of the waiting time before a complete connection is set up through the network is:

$$T_{max} = (N-1)T \tag{2}$$

where T is the maximum value for the time required for performing a data exchange, once the connection is established.

__Proof__. Let consider a request 1 blocked at stage i by a request K. Let $W_i^{(1)}$ be the waiting time of the request 1 before the trunk at stage i is obtained; it can be expressed, in the worst case by the following formula:

$$W_i^{(1)} = T + \sum_{m=i+1}^{n} W_m^{(K)} \tag{3}$$

where T is the time spent by the request K for transmitting data, once the connection it set up, while the second addend in (3) is the sum of the time spent by the request K, waiting for the availability of each trunk of its path from stage i+1 to stage n. In the worst case, the request K can be blocked at a stage j, $i < j \leq n$, and the corresponding waiting time can be expressed by an equation similar to (3). Since all the requests have the same characteristics, the worst case value of $W_p^{(q)}$ is independent on q, therefore, the superscript can be omitted. The following recursive procedure can used for evaluating W_i:

$$W_i = T + \sum_{m=i+1}^{n} W_m \quad , \quad W_n = T \tag{4}$$

In the worst case, a request can be blocked at each stage, hence the total waiting time is the sum of the waiting times at each stage and it will be:

$$T_{max} = \sum_{r=1}^{n} W_r = (2^n - 1) \, T = (N-1) \, T \tag{5}$$

It is worth noting that a worst-case occurs when all the N input devices contemporaneously try to accede to the same output device; using a non-preemptive policy, there is one input device served last; it must wait until the other N-1 devices complete their transmission. Hence the result shown in (5) is the best possible for T_{max}.

Fig.8 shows a sample of the implementation of the WT policy in a delta network. Each 2x2 crossbar (SW) is associated with a control unit (CU), which decides the state of SW using the control signal S. Each control unit has two request inputs (RIN1 and RIN2), which notify CU that a request is pending at input 1 or 2. Each one of the AD1 and AD2 inputs is tied to the appropriate information line carrying the bit of the routing tag associated to the pending request. The ROUT1 and ROUT2 outputs are used for notifying the crossbar in the next stage that a message has been routed to it, therefore they are connected to the request inputs of the two successive crossbars. Obviously, the information and request lines are connected to the same pair of CU and SW.

The simple asynchronous sequential circuit shown in Fig. 9 implements the control unit of a 2x2 crossbar switch, for the WT policy. Since this sequential circuit evolves from a steady state directly to another, the delay due to this implementation of the CU is small.

3.3.2 Release And Re-try policy

In section 3.3.1, it has been shown that, when the WT policy is adopted, the trunks held by blocked requests cannot be used by other incoming requests for building their paths. Hence, both the trunks involved in a data exchange and the trunks unused, but held by blocked requests, are unavailable for building paths for the arriving requests. This fact suggests a more complicated policy for processing blocked requests, which will be referred to as the Release-And-try (RAR) policy. In using the RAR policy, we seek to decrease the number of the occupied trunks, by making the trunks occupied by the previously blocked requests available for the requests in arrival. To do this, when a request is blocked, the trunks previously occupied are released and the subset of the path already built is destroyed. Since the blocked requests cannot be lost, but anyway they must built the whole connection, after the trunks occupied are released, the input device waits for a time Z, then it reissues the request for the same connection. In comparing the RAR and the WT policies, it is worth noting that the number of trunks available for an arriving request is greater in RAR than in WT policy; on the other hand, the arrival frequency of the requests is larger in RAR than in WT policy, since when the RAR policy is used the arriving requests are those generated by the input devices plus those due to the blocked requests.

A crucial issue in the implementation of the RAR policy is the value of z. In fact, short waiting intervals cause a high arrival frequency due to the blocked requests, which reenter the network; while long waiting intervals cause useless waste of time.

Fig.7 may also be used as block diagram of a sample of a network managed by the RAR policy. Unlike the implementation shown in Fig. 9, some additional input and output signals for the CUs appear; their use can be explained by means of an

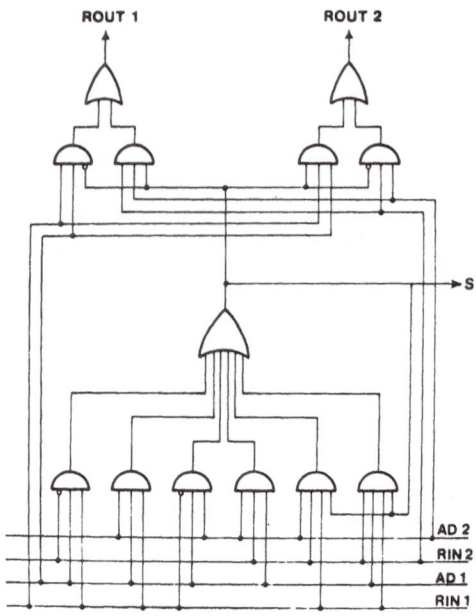

Fig. 9. Circuit implementing the control unit for WT protocol.

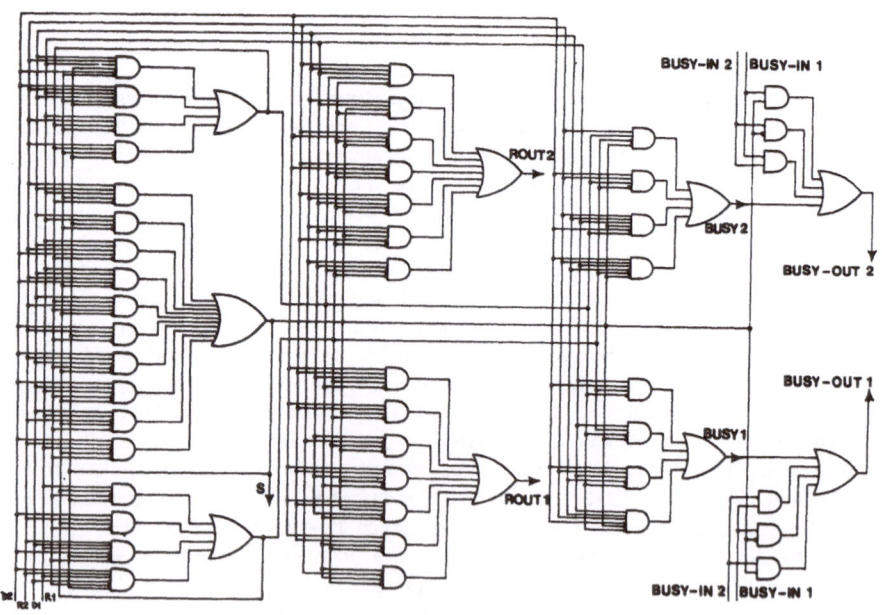

Fig. 10. Circuit implementing one control unit for RAR protocol.

example. Suppose that a request arrives at the input 1 of the crossbar i, the input $RIN1_i$ of CU_i is activated and the value of $AD1_i$ selects the output 2 of the same crossbar. Since the requested output trunk is free, the message is routed through SW_i to SW_k; contemporaneously, the ROUT 2 signal is activated so that CU_k is informed of the arriving message. Suppose that the output trunk requested at crossbar k is busy, the CU_k activates $BUSY-OUT1_k$, signaling the busy condition to CU_i. The latter repeats the same operation, upon $BUSY-IN2_i$ goes active, and so on. When the input device receives the busy signal from the CU in the first stage, it clears the request; this causes the clearing of every request signal in the path. When the $RIN1_k$ signal goes inactive, the CU_k replies, clearing the $BUSY-OUT1_k$ signal. This operation is repeated at each stage up to the input device. Only after the BUSY-IN signal is cleared, is the corresponding trunk released and made avilable for other requests.

Fig.10 shows an implementation of a CU which uses the RAR policy. This circuit is more complex that the corresponding one presented for WT; however, it is cheap enough to integrate several of these implementations in a single chip, as suggested in /13/.

4. Network implementation

This section deals with the problem of selecting a subset of a digit controlled networks built by 2x2 crossbar switches, suitable for LSI implementation. The objective of this selection process is to minimize the number of IC required for implementing the given network. The problem can be formally stated as follows:

$$\min C\ (\Omega) \qquad (6)$$

subject to:
$$P(\Omega) \leq Po \qquad (7)$$
$$A(\Omega) \leq Ao \qquad (8)$$
$$\Omega \in S \qquad (9)$$

where:

S = is the set of the possible (considered) LSI implementations of the elementary blocks.

Ω = is an element of S.

$C\ (\Omega)$ = is the number of packages of type required for implementing the given network.

$P\ (\Omega)$ = is the number of pins required by the implementation Ω

Po = is the maximum number of pins allowed.

$A\ (\Omega)$ = is a measure of the complexity of the implementation Ω

Ao = is the maximum value of A ( ) allowed by the current integration technology.

The first step is to find a common basic builing block for a class of digit controlled networks, as large as possible. In the paper of Lawrie /8/ it is pointed out that omega network can be built connecting, in a suitable way, other omega networks smaller than the given one.

On the other hand, Siegel in /14/ shows the equivalence between the networks of Pease /9/ and Lawrie, and Wu and Feng in /10/ state the topological equivalence between a baseline network and the simplified manipulator, flip, omega, reverse baseline and indirect binary n-cube networks. From the previous discussion, one may deduce that every one of the previoulsy mentioned networks can be viewed as composed by a set of omega (or indirect binary n-cube, or baseline, or reverse baseline etc.) networks, connected in a suitable way.

For example Fig.11 shows a 16x16 indirect binary n-cube network, built with eight 4x4 omega networks. Therefore, in the rest of the paper, the nxn omega network

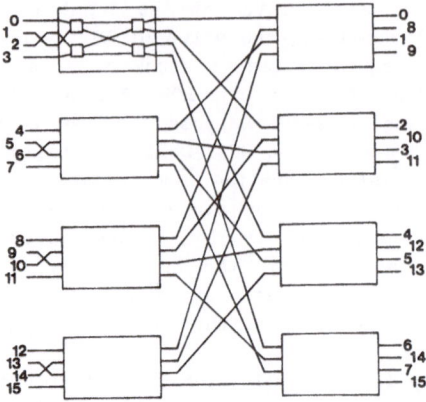

Fig. 11.   16x16 indirect binary cube implemented using
4x4 omega networks.

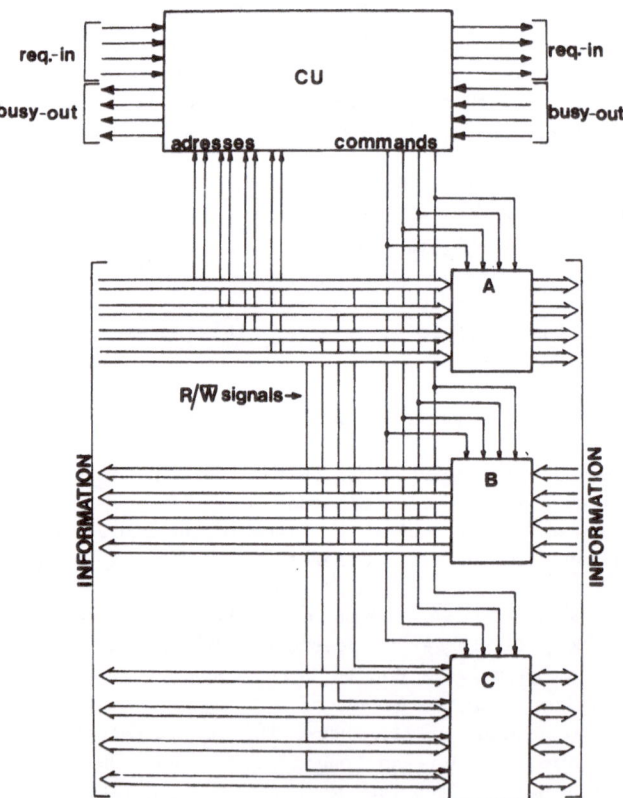

Fig. 12.   Block diagram of the implementation of each omega
network.

will be considered as the basic building block. Furthermore, it should be taken into account that each device, connected to the network, requires the parallel trasmission of data, address (or multiplexed data/address) and control signals, that is, for every input (output) device, w inlets (outlets) are required. Thus, it is possible to define S as the set of nxn omega networks which allows the parallel trassmission of w bits per device connected; where $n = 2^q$, q and w are integers. Any element of the set S will be referred to as Ω (w, n).

Let X_{ik} (i = 0, 1, ..., n -1 ; k= 0,1, ... w-1) one inlet of Ω (w,n) , and Y_{jk} (j= 0,1, ..., n-1;k= 0,1...,w-1) one outlet of Ω (w, n), it can be easily shown that:

$$X_{jk} = \bigvee_{i=0}^{n-1} (x_{jk} \wedge z_{ij})$$ (7)

Where Zij is a boolean function assuming the value "true" when the status of the network allows the connection between the devices i and j, "false" otherwise. The status of the network is defined by a set of control signals $\{s_i\}$; a Ω (w, n) block requires one control signal every one 2x2 crossbar, that is $\underline{n/2} \lg_2 n$ control signals.

It will be assumed that every control signal assumes the value "true" for direct connection and the value "false" for the exchange connection.
Since, in digit controlled networks only one route between every pair of devices exists, only one control signal per network stage is involved in any connection. Therefore, the function Z_{ij} assumes the following simple form:

$$Z_{ij} = \bigwedge_{K=1}^{\frac{n}{2} \lg_2 n} b_{k_{ij}}$$ (8)

where $b_{k_{ij}}$ assumes the value "true" when the control signal involved in the i-j connection at stage k assumes the value allowing that connection, and the value "false" otherwise.
For example, the function Z_{53} assumes, in the network shown in Fig.12, the following form:

$$Z_{53} = \bar{s}_2 \wedge \bar{s}_7 \wedge \bar{s}_{10}$$ (9)

It can be easily recognized that Y_{jk} may be expressed as the disjunction of n implicants. These implicants are essential, because a different variable (X_{ik}) appears in every term of the disjunction. Thus, the boolean function Y_{jk} must be implemented, usign 2 logic levels, by at least n+1 gates. Since in Ω(w, n) the function Y_{jk} must be replicated nw times, the following formula gives the total number of gates required for implementing the switching function for one (w,n):

$$A_1\{\Omega (w, n)\} = G_1 (w, n) = wn (n+1)$$ (10)

This equation is valid if one assumes that the signals are always transmitted in the same direction. In effect, data lines must be bidirectional in processor-memory interconnections. In this case, it is possible to realize a bidirectional basic block doubling the network defined for the unidirectional signals and using three-state outputs for such switching networks. Therefore, for one bidirectional Ω(w, n) the number of gates required may be expressed as follows:

$$A_2\left\{\Omega(w,\ n)\right\} = G_2\ (w,\ n) = 2\ w\ n\ (n+1) \tag{11}$$

The equations (10) and (11) give the number of gates required for implementing the switching function, using two gate levels. This same function may be implemented using a number of gate levels greater than two, in this case, both complexity and speed decrease. In fact, implementing one (w,n) using $\log_2 n$ gate levels, the gate count is equal to $1.5wn\ \log_2 n$, for unidirectional blocks, $3wn\ \lg_2 n$, for bidirectional ones. As will be shown later, increasing the values of n and w, the pins available are saturated when chip area is still available, therefore time saving is more important than complexity saving. In this paper, only the two gate level implementation will be considered.

The second feature of an IC implementation of $\Omega(w,n)$ is the number of pins required. This number is the sum of the following terms:
1) the number of external connections due to the inlets and the outlets of $\Omega(w,n)$;
2) the number of control signals required by $\Omega(w,n)$;
3) the number of connections for power supply:
4) the number of signals required for deciding the direction of data transfer (for bidirectional blocks only).

The first term may be deduced by the definition of $\Omega(w,n)$ and it is equal to 2wn. The second term, as indicated above, is equal to $0.5n\lg_2 n$; 2 pins are considered for power supply. Thus, the number of pins required by the LSI implementation of one unidirectional $\Omega(w,n)$ is given by the following equation:

$$P_1\left\{\Omega(w,n)\right\} = L_1\ (w,n) = 2\ wn + \frac{n}{2}\ \lg_2\ n+2 \tag{12}$$

For bidirectional $\Omega(w,n)$ n more leads are necessary to decide the direction of the data transfer so, in this case, the total number of pins can be expressed by:

$$P_2\left\{\Omega(w,n)\right\} = L_2\ (w,n)=(2\ w+1)\ n+\frac{n}{2}\lg_2\ n+2 \tag{13}$$

4.1 Control Scheme

Since the ratio gates/pins of the IC proposed in the previous section is very small, one might think that it is feasible to put in the same chip both the connecting subnetwork and its control unit. The latter needs a set of input and output signals (request, busy signals,...), therefore many other pins are required. A more attractive solution is depicted in Fig.12, the control of a $\Omega(w,n)$, built with the ICs proposed in the previous section, is concentrated in a dedicated chip.
The mechanism of searching and allocating the path requested through the network is described below. The request generated by a processor is issued at the input to the control unit of the $\Omega\ (w,n)$ in the first stage connected with that processor; each request is issued with the binary output device address. The control unit in the first stage receives the request signal and $\log_2$ bits of the output device address.
This set of $\log_2 n$ address bits in chosen on the basis of the type of network implemented. In an omega network, for instance, the most significant $\log_2 n$ bits are connected with the control unit of the first stage subnetwork, the next $\log_2 n$

most significant bits are connected with the second stage control unit and so on:
On the basis of the state of the switching elements, the active requests and the addresses related to them, the control unit decides whether or not to accept the request. If the request for the second stage is generated and the status of the switching elements is changed to accomodate the new connection.

When the second stage receives the request issued by the first stage, an analogous mechanism starts. Thus, the path requested is searched for and allocated, stage by stage, until the target outlet is reached.
If, at any stage, the control unit detects a conflict between the requested path and the connections active at that time, the status of switching elements is not changed and a busy signal is issued back to the processor through the previously allocated connections. When the busy signal is received by the requesting processor, the associated request is turned off and reissued later.

The connections are kept until the processor, which issued the request, terminates the transfer of information; at that time, it clears the request and releases, stage by stage, all the trunks which compose the whole connection.
An integrated circuit implementing the above described protocol for a Ω (w, n) requires: n input signals for the path requests coming to the next stage; n inputs for busy signals, arriving from the next stage; n outputs for busy signals, the previous stage; n $\lg_2 n$ inputs which define the status of the controlled switching elments.
Therefore, the total pin count for this control element is given by the following formula:

$$M (n) = 4 n + \frac{3}{2} n \lg_2 n + 2 \tag{14}$$

Where 2 pins are considered for power supply.
Evaluating (12) for several values of n, it may be noted that the implementations with $n = 2^q > 8$ are unfeasible, if we assume that the maximum number of pins allowed by the current technology is 120.
In order to solve the problems defined by (5), (6), (7) and (8), the analytical form of $C (\Omega)$, required for implementing a network of the class considered in this paper is the sum of four terms: the number of chips required to transmit signals in one direction; the number of chips required to transmit signals in the other direction; the number of chips required to transmit bidirectional signals; the number of chips performing control functions.
Let:
N = the number of processors equal to the number of memory banks.
C_1 = the number of signals issued by each processor to the network.
C_2 = the number of response signals issued by each memory bank to the network.
D = the number of bidirectional signals exchanged between each processor—memory pair.
It should be noted that a NxN network of the class considered here may be built with $m = \log_n N$ stages of N/n nxn omega networks allowing the transmission of the C_1, C_2 and D signals. One of such omega networks requires:

$$\left\lceil \frac{C_1}{W_1} \right\rceil + \left\lceil \frac{C_2}{W_1} \right\rceil \tag{15}$$

chips for the transmission of unidirectional signals, if the LSI implementation of a unidirectional Ω (w_1, n) is used as the basic block; $\lceil D/w_2 \rceil$ chips for the transmission of bidirectional signals, if the LSI implementation of a bidirectional Ω (w_2, n) is used as the basic block.

Then the total chip count becomes:

$$C\ (\Omega) = (1 + \left\lceil \frac{C_1}{w_1} \right\rceil + \left\lceil \frac{C_2}{w_2} \right\rceil + \left\lceil \frac{D}{w_2} \right\rceil)\ \frac{N}{n}\ lg_n N \tag{16}$$

However, a subset of the C_1 signals is issued to the control units in order to select the path requested; therefore, these signals might not travel further on, when the appropriate control unit is reached. Then, C is given by the following:

$$C\ (\Omega) = \frac{N}{n} \sum_{h=1}^{m} (1 + \left\lceil \frac{C_2}{w_1} \right\rceil + \left\lceil \frac{D}{w_2} \right\rceil + \left\lceil \frac{C_1 - min\ (hlg_2 n,\ lg_2 N)}{w_1} \right\rceil) \tag{17}$$

For example, for N= 64, C_1=14, C_2=2 D=16 and Po=40, we obtain that 512 packages are required for implementing the network; that is, only 8 packages per processor.

5. Multiple path networks

Since interconnection networks are intended for large multiprocessor systems, fault-tolerance issues are of primary importance. Indeed, the system failure rate is tightly related to its complexity, since, in general, it is roughly equal to the sum of the failure rates of the components.

Fault-tolerance can be introduced in an interconnection network by using one or more techniques, which can basically amened to one of the following three classes:

- use of error correcting/detecting codes for transmitting data through the network;
- use of the intrinsic redundancy of a multiprocessor system for reassigning the tasks to the different processors, so that the faulty subsets of the network are no longer used;
- use of another class of networks with more than one path between each input-output pair, still conserving a cost O(NlogN); the multiple paths are used to circumvent faulty subsets of the network.

In the rest of this section, only networks belonging to the third class will be discussed, since the other two techniques are more related to code theory and fault-tolerant operating systems, respectively.

The simplest way for obtaining multiple paths is to add on extra stage to a single path network. This technique has been applied to a binary cube network /15/. In the original network, which belongs to the class of delta, the routing data is computed as $t_j = s_{n-1-j} \oplus d_{n-1-j}$ ($0 \le j \le n-1$), then the last bit of the tag is used to control the first stage and so on. In other words, if the source and destination nodes differ in the first bit, the most significant bit of T must be 1.

The extra stage is placed at the network inputs, and its switches are laid out to the original network so that if its 2x2 crossbar switches are set at X, the least significant bit of the destination routed through the original network according to T is complemented, viceversa if the extra stage is set at T.

Hence, the additional stage is able to perform the same routing operation of the last stage of the original cube network it turns out that, if $s_o \oplus d_o = 1$, it is possible to reach the right destination either setting the extra stage at X and the last stage at T or viceversa; analogously, if $s_o \oplus d_o = o$, the extra and the last stages must be set at the same state.

With the extra stage it is possible to have 2 possible paths between each input-output pair; the additional cost is that of one stage for a network composed of $\log_2 N$ stages. On the other hand, an a priori decision on which path should be followed is needed; indeed, once the message has traversed the extra stage, there is only one path for reaching the destination.

Another class of multiple path networks recently introduced allows a message to be rerouted at each step of the routing algorithm, according with the faulty, non-faulty state of the switches, such a technique is referred to as dynamic rerouting, and it allows also to achieve a performance improvement since the message may be rerouted on-the-fly not only when a fault is encountered, but also when a switch is busy because another connection is in progress.

Usually, these multiple path networks are described, by means of a graph, which is composed by $1 + \log_2 N$ columns of N nodes. Column 0 represents the network inputs, colunum N represents the network outputs and the other columns are switches the nodes are connected by a set of edges representing the links.

The first example of dynamic rerouting has been presented for the IADM network /15/. This network uses 3 input and 3 output links for each node in an internal colunum. Hence, when a message reach a node, it needs a three valued tag digit for selecting the output link. Therefore, the tag T is composed of $\log_2 N$ 3-valued digits; since T=D-S, in general, several patterns of $\log_2 N$ 3-valued digits may be used to represent $0 \le D-S \le N$. Each value of a digit selects a different node output hence the different patterns leads to different paths.

However, using a number representation with 3-valued digits (-1,0,1), the value 0 may be represented by only one combination constituted by all'0s. If multiple paths are needed for every value of T, it is necessary to switch to network based on nodes with 4 input and output links. One of such networks is the F /16/; if the nodes in the same column are numbered from 0 to N-1, and the colunms are numbered from 0 to $\log_2 N$, each node $P=(P_{n-1},...,P_o)$ in colunum j ($0 \le j < \log_2 N$) is connected to the following four nodes in colunum j+1:

$$(P_{n-1}, \ldots, P_{j+1}, P_j, \ldots, P_o) \tag{18}$$

$$(P_{n-1}, \ldots, P_{j+1}, \overline{P}_j, \ldots, P_o)$$

$$(\overline{P}_{n-1}, \ldots, \overline{P}_{j+1}, P_j, \ldots, P_o)$$

$$(\overline{P}_{n-1}, \ldots, \overline{P}_{j+1}, \overline{P}_j, \ldots, P_o)$$

The final result is shown in Fig. 13 for N=8.

The main feature of the F-network is the ability to reroute the message at each node except those in colunum $(\log_2 N)-1$, since 2 of the output links may be always used to build a path for the same destination, whatever source-destination pair is selected.

An additional improvement may be attained if at each node it is possible to know the state of its successors on the next two columnus. In /17/ , for example, modifications of the IADM network are presented, which allow the network to recover up to 2 or 3 node or link failures, according to whether 4x4 or 5x5 nodes are used.

It could seem that only a moderate fault-tolerance improvement is obtained by using multiple path networks, while the cost is roughly doubled. However, the figures shown above for the modified IADM are only the worst-case; in general, multiple path networks allow to recover from a larger number of faults than in the worst-case. An analysis of the MTBF has been performed on the F network, considering only node failures. The results of this analysis are shown in Fig. 14 where the MTBF vs network size of the F compared with that of a system with h delta networks in parallel.

It can be seen that the multiple path network achieves an MTBF two orders of magnitude better than parallel single path networks.

Concluding remarks

With the continuous dramatic decrease of the cost and the increase of the performances of the integrated circuits, multiprocessor systems composed by a very large number of processor and memory modules become even more appealing. In these systems, a primary role is played by the communication subsystem, since a careful design is necessary to exploit the potential computation power of the processors used.

A class of communication networks for MIMD systems has seen considered in this paper. The networks in this class allow to establish dynamically the connections between every pair of devices; their complexity is O(NlogN), which is larger than in single shared bus configuration, and it is significantly less than full crossbar switch.

These networks allow more than one transmission to be carried out in parallel, and though some conflict condition may occur, their bandwidth approaches that of a full crossbar switch.

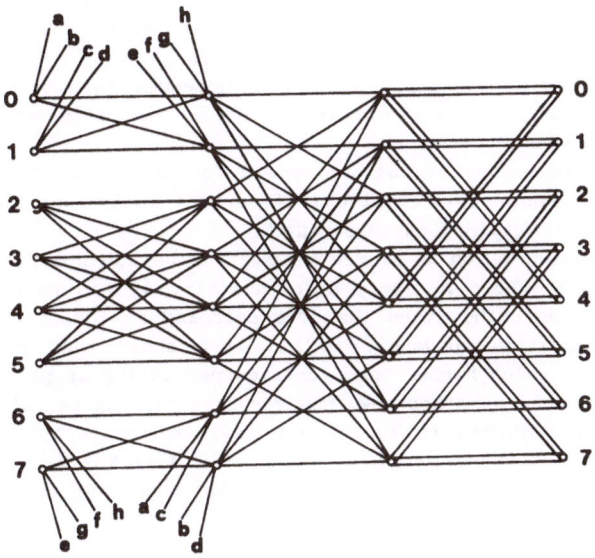

Fig. 13   8x8 F-network.

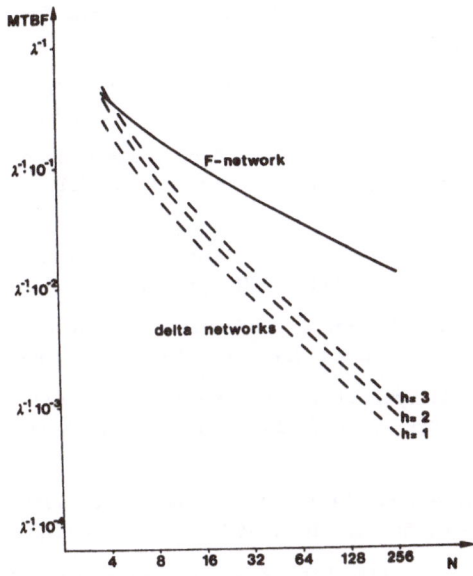

Fig. 14. MTBF vs. network size for the F and h
parallel delta networks.

Within this class of interconnection networks, both single and multiple path networks exist. A typical representative of the former ones is the set of delta networks, whose definition has been recalled in this paper.

Two important implementation issues of delta networks have been discussed: the data exchange protocol, and the definition of ad hoc integrated circuits for implementing the network.

A wide spectrum of solutions has been proposed for the former issues, they range from packet to different implementations of circuit switching; the complexity of each solution has been outlined in this paper.

Putting entire subnetworks on a single chip gives us the possibility of a cheap implementation of the whole network; unfortunately, it has been shown that the limited number of pins rather than the area occupied is the strongest constraint for an LSI implementation. However, it has been shown that the cost of this implementation is not which higher than for a corresponding single shared bus.

Multiple path networks are rather new, however several proposal already exist; the goal in using these networks is to achieve fault-tolerance by providing several way for implementing a connection. A secondary objective may be the improvement of performances, through dynamic rerouting. Networks achieving both objectives have been discussed.

References

1. J.A. Harris and D.R. Smith "Hierarchical Multiprocessor Organization", "Proc. Fourth Symp. Computer Architecture, Mar. 1977, pp. 11-18.

2. G.H. Barnes ed al., "The Illiac IV Computer", IEEE Trans. Computers, Vol. C-17, No. 8 Aug. 1968, pp. 746-757.

3. H.T. Kung, "The Structure of Parallel Algorithms", in Advances in Computers, Vol. 19, M.C. Yovits, ed., Academic Press. N.Y.? 1980.

4. F.P. Preparata and J. Vuillemin. "The Cube-Connected Cycles: A Versatile Network for Parallel Computation." Comm. ACM, Vol. 24, No. 5, May 1981. pp. 300-309.

5. H.S. Stone. "Parallel Processing with the Perfect Shuffle, "IEEE Trans. Computers, Vol. C-20, No. 2 Feb. 1971, pp. 153-161.

6. J.H. Patel, "Processor-Memory Interconnections for Multiprocessors "Proc. Sixth Annual Symp. Computer Architecture, Apr. 1979, pp. 168-177.

7. T. Feng, "Data Manipulating Functions in Parallel Processors and Their Implementations". IEEE Trans. Computers, Vol. C-23, No. 3, Mar. 1974, pp. 309-318.

8. D.H. Lawrie, "Access and Alignment of Data in an Array Processor" IEEE Trans. Computers, Vol. C-24, No. 12, Dec. 1975, pp. 1145-1155.

9. M.C. Pease, "The Indirect Binary n-Cube Microprocessor Array", IEEE Trans. Computers, Vol. C-26, No. 5 May 1977, pp. 548-573.

10 C.Wu and T. Feng, "The Reverse-Exchange Interconnection Network", IEEE Trans. Computers, Vol. C-29, No. 9 Sept. 1980, pp. 801-811.

11 L.R. Goke and G.J. Lipovski, "Banyan Networks for Partitioning Multiprocessing Systems", Proc. First Annual Computer Architecture Conf., Dec. 1973, pp. 21-28.

12 Ciminiera L., Serra A., "Modular Interconnection Networks with Asynchronous Control", Proc. 14th Hawaii Intern. Conf. on System Sciences, Jan. 1981, pp. 210-218.

13 Adam III G.B., Siegel H.J., "The Extra Stage Cube: a Fault-Tolerant Interconnection Network for Supersystems", IEEE Trans. Computers, vol. C-31, n. 5, May 1982, pp.

14 R.J. McMillen and H.J. Siegel, "MIMD Machine Communication Using the Augmented Data Manipulator Network", Proc. Seventh Symp. Computer Architecture June 1980, pp. 51-58

15 Ciminiera L., Serra A. "A Fault-Tolerant Connecting Network for Multiprocessor Systems", Proc. 1982 Intern. Conf. on Parallel Processing, Aug. 1982, pp. 113-122.

16 Mc Millen R.J., Siegel H.J., "Performance and Fault-Tolerance Improvements in the Inverse Augmented Data Manipulator Network", Proc. 9th. Ann. Sym. on Computer Architecture, April 1982, pp. 63-72.

THE ORGANIZATION OF

PARALLEL PROCESSING MACHINES

A.M. Wood
Dept. of Physics
University College London
Gower Street
London WC1

1 The Classification of Parallelism

A computing system, in the widest sense, is composed of many units
each of which can be in one of a set of states, and which may change
their states according to variations in their inputs. Since there
will generally be some sets of units which have no direct links
between them, the state transitions of these sets can occur simultaneously and independently, and thus in parallel. It is always possible to view a system at a level at which this is true and thus all
systems exhibit parallelism.

Conversely, for a system to be of any practical use, some causal
links between units must be present. Causality implies an ordering of
actions in time, which in turn implies serialism. Therefore, all systems have both serial and parallel aspects.

In practice however, we are interested in systems at various levels
of abstraction, and thus our view as to the relative importances of
the serial or parallel aspects will vary: as the level of abstraction
increases, the more "parallel" a system becomes. For instance, at
the level of 'bits', a Turing machine may be regarded as more serial
than a common microprocessor, both of which would be less parallel
than a brain. If a physicist were to look at these systems however,
he may only see the interactions of fundamental particles, making a
parallel adder look extremely serial.

It is thus very important to make clear the level of abstraction being used when discussing parallel machines.

Several attempts have been made to suggest ways of expressing this abstraction in a convenient and simple way [2,3], a recent review being Cantoni [4]. The most frequently used scheme is that due to Flynn [1], where he gives three separate criteria by which computer architectures may be classified. The first is to note the number of 'data' and 'instruction' streams that the machine can handle simultaneously. This leads to a simple notation in which the initials "S" and "M" stand for single and multiple, and the initiials "I" and "D" stand for instruction- and data-streams respectively. Thus for instance, the string SIMD, would mean single instruction-, multiple data-stream.

The second 'axis' of the system is a description of the interconnection between storage and processor(s). This takes the form of up to three matrices describing the storage-storage, processor-storage and processor-processor transfer times. These are related to the dimensionality and connectivity (topology) of the processors and storage.

The final part of this classification scheme is a measure of the degree by which an instruction stream may overlap its separate operations. Flynn calls this the "inertia factor" and it is most commonly encountered in conventional computers where the execution of one instruction is overlapped with the decoding of the next.

Unfortunately, Flynn's scheme is generally misused in the literature, which has reduced its potential as a formal classification procedure. This misuse occurs in two ways. Firstly only the first (S,I,M,D) 'axis' of the system is commonly used, with no regard to the vital points of connectivity and dimensionality. Secondly, it is rare that the units of data and instructions are defined, thus causing potential inconsistencies in inter-architecture comparisons. Hence Flynn's scheme, as it is (mis)used, is little more than four mnemonics (SISD, SIMD, MISD, MIMD) which help to trigger associations in the minds of people already familiar with the field. In this way it is marginally less useful than informal classification by the description of a machine in English or some other language. However, due to its wide use, it is appropriate that some examples be given in order that the associations mentioned above may be initialised. The next

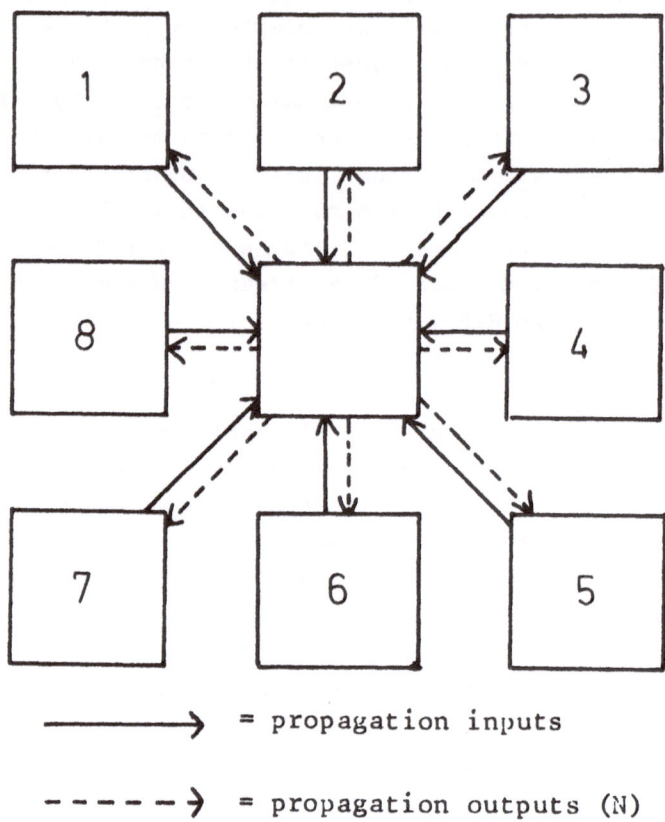

$\longrightarrow$ = propagation inputs

$- - - - \rightarrow$ = propagation outputs (N)

Figure 1   CLIP Neighbourhood Interconnections

section will describe some existing parallel machines which may considered typical of the architectures currently under investigation,
while indicating how the Flynn mnemonics are conventionally applied.

2   Parallel Hardware

Bearing in mind the qualifications expressed in the preceding section, the Flynn mnemonics give rise to four groupings of computer
architectures thus:

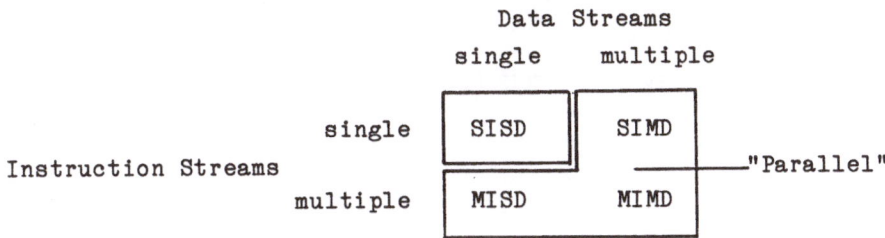

According to this, the appearance of "M" in the mnemonic indicates
that parallelism exists when the system is viewed at the level at
which the instructions or data are defined. Thus, if the data unit
were the byte, an 8-bit micro could be regarded as SISD, whereas a
VAX (32-bit machine) could be classed as SIMD since an operation
such as stripping the parity bit from four ASCII characters (clearing the top bit of a byte) would take four instructions on the micro
but one instruction on the VAX. If the unit of data were the bit,
then both machines could be described as SIMD. The situation becomes
even less clear when the machines in question are re-configurable
during the course of a program, as is often the case (e.g. CLIP4).

2.1 Array Processor: CLIP4

CLIP4 [5,6] is an example of an <u>array processor</u>. That is its processing unit is a two-dimensional array of (96x96) boolean processors,
each of which is connected to its 8 nearest neighbours (fig. 1).
Each processor executes the same instruction at each step of a

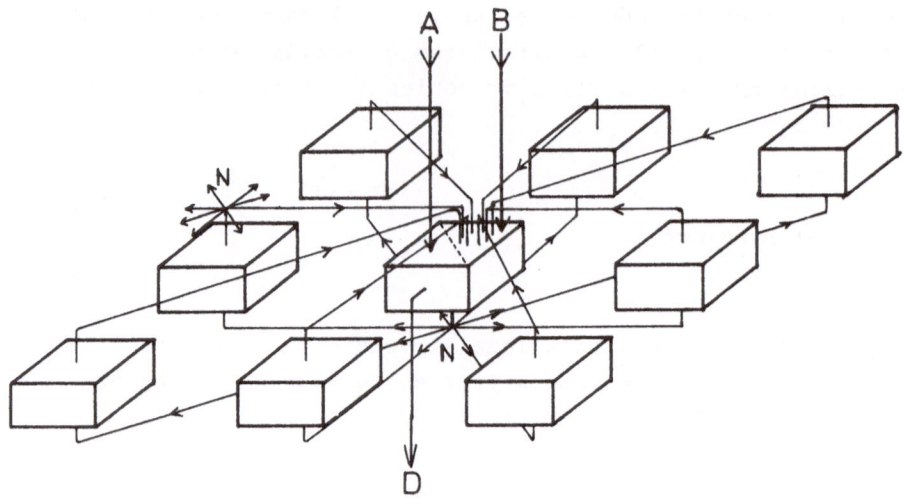

Figure 2   CLIP Data Paths

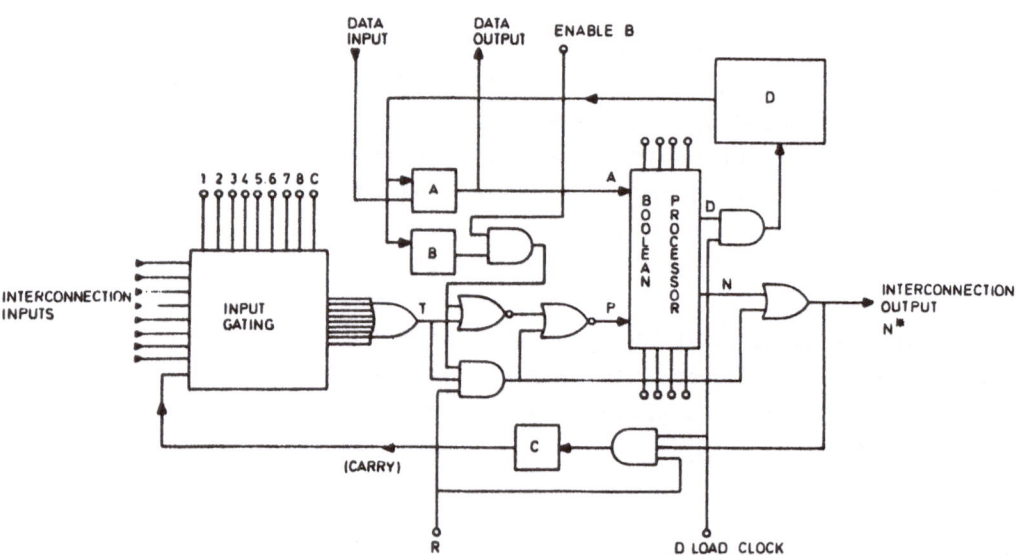

Figure 3   CLIP PE Internal Structure

program, but operates on its own local data inputs.

2.1.1 Classification

The nature of the parallelism on data combined with the use of a single instruction stream broadcast to every processing element means that CLIP is a 2-dimensional SIMD processor when the unit of data is the bit.

The set of neighbours from which a processing element (PE) receives data can be chosen from any subset of its eight surrounding processors, although each PE must have the same relative neighbours selected. It is therefore possible to configure the array (during the program) so that, for example, it looks like a 96-element list of 96-bit integers, all of which may be acted upon in parallel. It is thus also SIMD on 1-dimensional data in the form of integers or bit-strings. A microprocessor would be SISD on integers.

The CLIP series of machines, of which CLIP4 is the most recent, are general-purpose computers which were originally developed for use in image-processing. If the data items are defined as _images_ or two-dimensional arrays, then CLIP becomes an SISD machine since it operates on these items one by one.

In practice however, array processors such as CLIP (e.g. MPP [7]) are referred to as SIMD with the implication that the data unit is the bit.

2.1.2 Organisation

Each PE is formed from _two_ identical processors: one producing the output function, and the other giving the neighbourhood function (fig. 2). Any boolean function of two inputs can be generated by these processors, allowing complete generality. The output function (D) is what is usually thought of as the 'result' of the instruction,

138

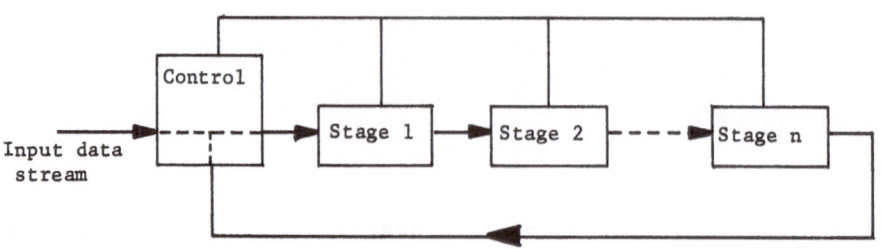

Figure 4   Basis of Cytocomputer

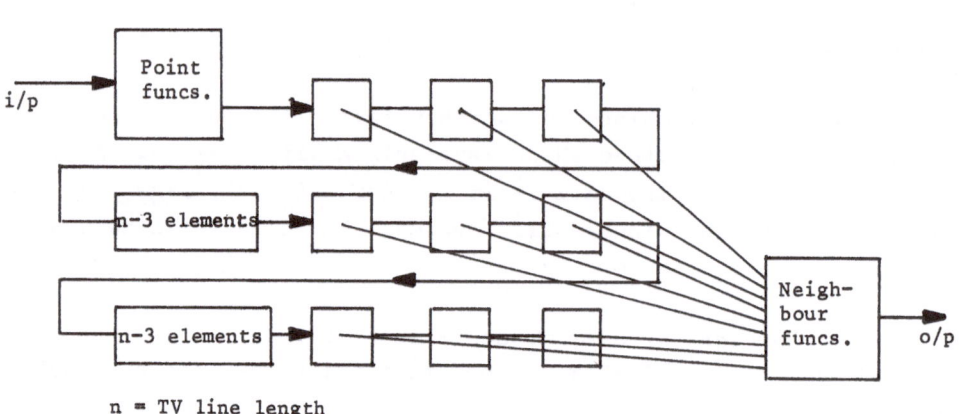

n = TV line length

Figure 5   Cytocomputer Stage

whereas the neighbourhood function (N) produces a signal to be sent to the element's neighbours. The structure of the CLIP4 processing element is shown in figure 3.

Three classes of array instruction can be identified. Firstly null propagation instructions are those in which no signal is sent to an element's neighbours, whose outputs are thus solely dependent on the one or two external input arrays. An example of the type of operation would be the logical AND of two binary matrices in parallel.

The second class of operation is local propagation where the output is a function of the element's input arrays and the states of its selected nearest neighbours. Such an operation would be the shifting of an array of data in one direction, or the detection of elements whose surrounds matched a certain pattern.

Finally, global propagation instructions occur when the signal sent to an element's neighbours is a function of the signals received from its neighbours. A recursive operation can thus occur in which information about an element's state can be transmitted globally across the entire array in one machine instruction. These powerful instructions have several uses, one common application in image processing programs being the extraction of the outer boundaries of objects in a binary scene, but ignoring any holes contained within the objects, again in a single array operation.

2.2 Pipeline: Cytocomputer

The ERIN Cytocomputer [8] is an example of a pipeline processor, this type of system being generally classed as MISD. The basic processing structure is as shown in figure 4 in which a stream of data enters the system, is processed by the first stage, the output from which is processed by the second stage, and so on. Thus parallelism exists since, as the pipeline fills, each stage is simultaneously operating on a different portion of the (partially processed) data stream. This parallelism is further enhanced when the length of the data stream is less than that of the pipeline, when the inactive stages "ahead" of the first element of data may be re-programmed "on the fly" while the data is circulating.

In fact, the Cytocomputer has a second layer of pipelining in each of the processing stages. Since it was designed to operate upon TV-like images, an array processor structure could be emulated using a 3x3 processor 'window' and delay shift-registers as shown in figure 5. As a serial data stream passes through a stage, it is equivalent to the window being scanned over the two-dimensional TV image.

Thus the two layers of pipelining can be seen to map into two levels of parallelism: the concurrent execution of different functions, and the partially parallel calculation of the functions themselves.

2.3 Data Flow: the Manchester Machine

There is currently much interest in computer structures which correspond to the data-flow model of the computational process. In this organisation, instead of specifying the way in which <u>control</u> flows through a program, as reflected in conventional programming languages, the <u>data</u> dependencies are stated; the execution of the various sub-functions in a program are controlled solely by the existence of their input values. In this way, any inherent parallelism in a program may be identified for potential concurrent execution on the hardware.

As an example, consider the following program fragment:

$$x = (a+b) * \sqrt{c}$$
$$y = (a+b)+3$$
$$z = x*y$$

This may be represented in terms of its data-flow graph shown in figure 6. It is clear that it is possible for $(a+b)$ and $\sqrt{c}$ to be calculated concurrently. Also if the square-root takes longer than the sum, y could be calculated while $\sqrt{c}$ is finishing, since both of its data values would be available. The final result, z, would be produced only when both x and y had been evaluated.

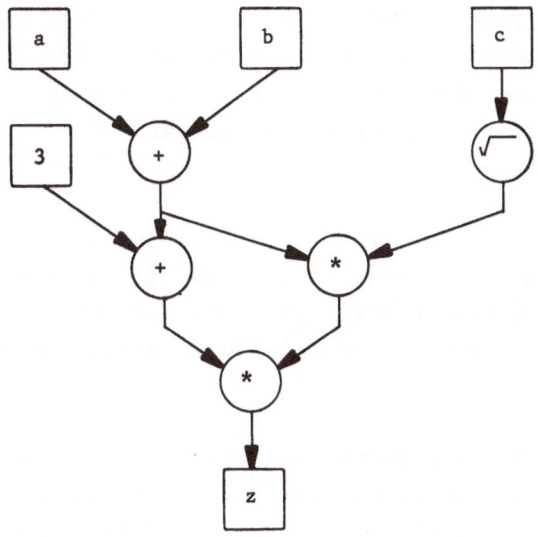

Figure 6   A Program Graph

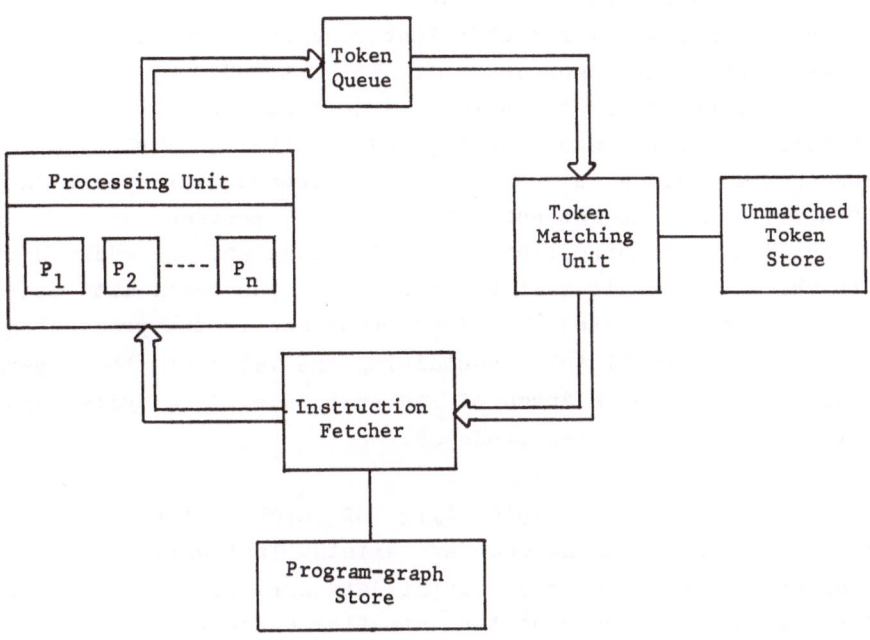

Figure 7   Manchester Data-flow Machine

The potential for parallelism is therefore easily seen. Additional parallelism is possible in the form of pipelining. Once data values have percolated down the program graph sufficiently, a new set of input values could be entered at the top and the idle nodes (processors) re-activated.

Since there can be many processors (nodes in the program's data flow graph) evaluating different functions in parallel on multiple streams of data (represented by arcs in the program graph), machines executing such a computational model are generally termed MIMD.

Various systems have been developed using the data-flow concept [e.g 9,10]. One such has been constructed at Manchester University by Watson and Gurd [11]. A simplified diagram of this machine is shown in figure 7, in which input/output has been omitted. The values that exist on the arcs of the program's data-flow graph are termed "tokens", and take the form of packets containing data values and control information, which flow around the communications ring. The control fields consist of an address representing the destination node (function) of that token, and a tag which allows the implementation of a mechanism for efficient recursion and iteration. When each token leaves the token queue it is matched, by the matching unit, to the other input token for its destination node. This matching is done on the basis of the tag and destination fields. The pair of tokens then passes to the instruction fetcher which uses the destination field and the representation of the program in the graph store to find the appropriate operation code and the address of the node to which the result must be directed. The data-data-operation packet is then directed to a free processor in the processing unit which produces a result token containing the value of the operation and the destination address of its next node. This cycle continues until the final results are produced.

The notable feature of this type of architecture is that no equivalent to the 'program counter' exists. Instructions are fetched from the program store as their input data arrives. This fundamental property is a consequence of the data-flow concept and is what gives rise to the realisation of parallelism in the multiple independent processors.

3 Summary

Concurrency in computer systems can exist either as data parallelism, or instruction parallelism, or both. The Flynn scheme uses this fact along with other information to attempt a taxonomy of processors. The common usage of this scheme however, is to consider <u>only</u> the data and instruction parallelisms to give four mnemonics which, while being of questionable value as a classification, have become almost universally accepted.

Three operational machines have been described which fall into the parallel groupings of the Flynn mnemonics. The array processor (CLIP4) uses parallelism at the bit level to obtain very high speed processing of two-dimensional data. The pipeline machine (Cytocomputer) contains restricted bit-level parallelsim in the 3x3 window processors, and restricted functional parallelism in the sequence of stages. The data flow machine exhibits full functional parallelism up to the limit of the relatively few (15) processors in the system.

4 References

[1] Flynn, M.J.
     "Some Computer Organisations and their Effectiveness"
     IEEE Trans. Comp. C-21 p948 1972

[2] Hockney, R.W. and Jesshope, C.R.
     "Parallel Computers"
     Adam Hilger
     1981

[3] Danielsson, P-E. and Levialdi, S.
     "Computer Architectures for Pictorial Information Systems."
     IEEE Computer 14 p53 1981

[4] Cantoni, V.
     "Classification Schemes for Image Processing"
     Proc. NATO ASI, Cetraro, 1983

[5] Duff, M.J.B.
     "CLIP4: A Large Scale Integrated Circuit Array
      Parallel Processor."
     Proc. 3rd. IJCPR p728 1976

[6] Wood, A.M.
     "The CLIP4 Array Processor"
     Proc. IEE Symp. on Real-time Processor Architectures...
     IEE London 1982

[7] Batcher, K.E.
     "Design of a Massively Parallel Processor."
     IEEE Trans. Comp. C-29 p83 1980

[8] Lougheed, R.M. et al
     "Cytocomputers: Architectures for Parallel
      Image Processing."
     Proc. IEEE Workshop on Picture Data Description
     and Management. p281 1980

[9] Dennis, J.B. et al
     "Building Blocks for Data Flow Prototypes."
     Proc. 7th Symp. Computer Arch. p1 1980

[10] Arvind and Gostelow, K.P.
      "A Multiple Processor Dataflow Machine that
       Supports Generalized Procedures."
      Proc. 8th Symp. Comp. Arch. p291 1981

[11] Watson, I. and Gurd, J.R.
      "A Practical Data Flow Computer."
      IEEE Computer 15 p51 1982

ORGANIZATION OF MULTI-PROCESSOR SYSTEMS FOR IMAGE PROCESSING

V. Cantoni

Dipartimento di Informatica e Sistemistica
Universita' di Pavia
Strada Nuova 106/c
27100 Pavia (Italy)

ABSTRACT

There has been an increasing interest in the analysis of image data with high-speed parallel hardware. The advent of VLSI technology spurred the construction of such systems and several practical commercial applications appeared in the late 70's. Much work has been done to develop parallel processors for low level image processing. However, in the image analysis field an highly effective solution, till now, has not been found. In this paper the nature of image processing tasks is outlined and the organization of the multiprocessor systems which have been developed for these tasks are reviewed.

1. INTRODUCTION

In many image processing applications, from robotic vision to the analysis of the earth's surface from satelite imagery, large amounts of computation to be done in a very short period of time, are frequently required. Conventional general purpose computers are unable to achieve these high computation rates. So, special parallel computer architectures, that have become feasible with the advent of VLSI technology, have been constructed, and others are currently being developed for these purposes.

Image processing problems may be broadly divided into two classes: low level and high level image processing (or image understanding, image interpretation or image analysis) (see Fig. 1). The former class deals with geometric correction, noise removal, restoration, edge detection, object segmentation, feature extraction, etc.. Such tasks possess the following fundamental features: i) the output has the same matrix size as the input; ii) identical operations are applied to all pixels; iii) the computation for a pixel requires only the pixels in the local neighborhood of that pixel. The latter class deals with classification of segments or feature into known classes using statistical methods, graph algorithms, syntax analysis, etc.. For these cases, the image is often no longer considered as a large matrix of pixels. A set of parametric measures or a relation graph can provide more convenient data structures to represent segments of the image. In these cases a set of sequential processes which may be conducted independently, in parallel are often required.

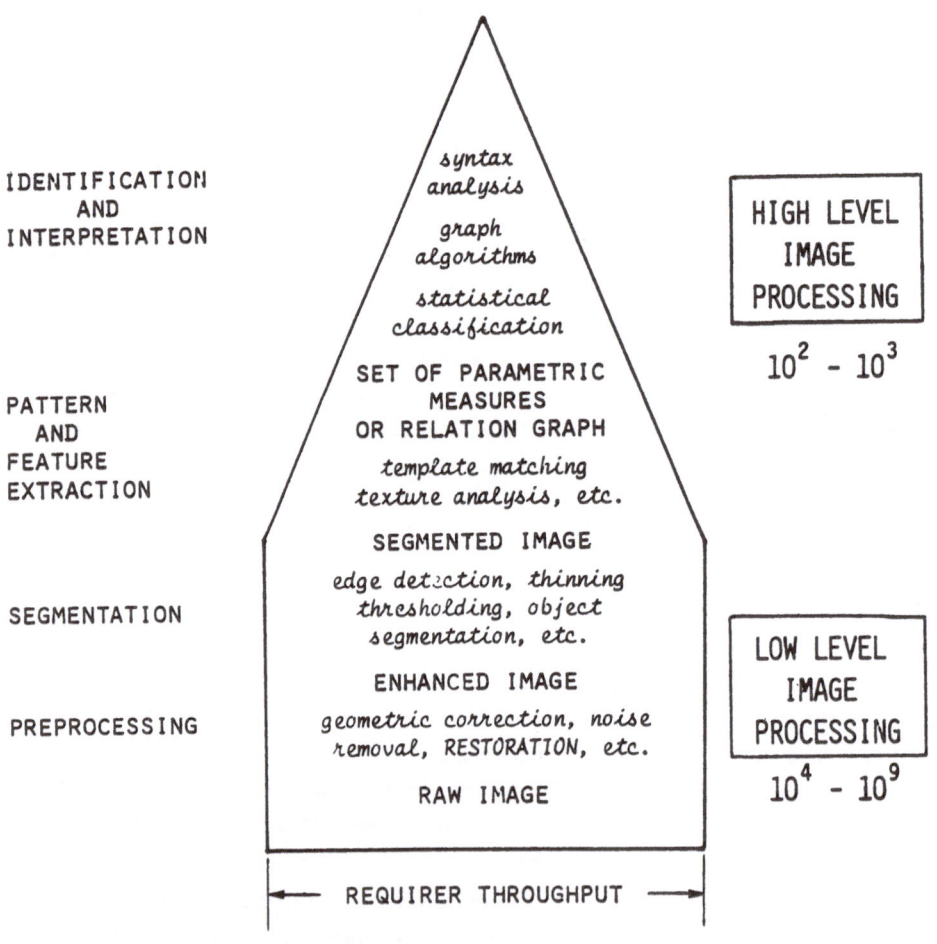

FiG 1 - Low level and high level processing stages of images.

In this connection a computer architecture designed for image processing should have as many of the following features as possible:

- a high replication of arithmetic/logic units, operating synchronously in the SIMD mode, within the limit of one ALU for each pixel in order to fully exploit the spatial parallelism of the data structure in low level processing.

- a high degree of pipelining to overlap instruction execution in order to exploit the parallelism of the computation structures, particularly the fact that the same set of operations is applied to all pixels, a high

- a multiprocessor, in the MIMD mode, architecture using suitable interconnection scheme, to permit asynchronous computation and communication between processors in order to support high level processing.

- either a core memory large enough to contain the image or a high speed transfer mechanism for I/O image operations.

In this paper the fundamental concepts of different parallel architectures for image processing will be described, highlighting the more recent design. Existing system architectures are here grouped in four classes: cellular arrays, pipeline architectures, bus oriented architectures, and architectures with reconfigurable interconnection schemes.

2. CELLULAR ARRAYS

A fundamental problem in the past has been to achieve the very high computation rates needed to implement low level image processing algorhythms. Highly parallel SIMD computer architectures have been proposed to solve these problems for over 20 years. One of the first proposals of this type was by Unger in 1958 [1]. This machine involves one control unit and a large number of ALU's with local memories, called processing elements (PE's); each PE performs the same function as all others but on different data elements. In the optimal situation, one PE is assigned to each pixel in the image, so that, the computation time only depends on the computational depth and not on the image size. At that time only discrete component technology was available and systems involving thousands of PE's could not reliably be built.

With the advent of VLSI technology it has become feasible to implement these proposals and, in fact, several such systems have already been developed. CLIP 4 developed at University College London [2] involves a matrix of 96 x 96 PE's and is based on an LSI chip containing 8 PE's (see Fig. 2). The Distributed Array Processor (DAP) is an operational system developed by International Computer Limited and involves a matrix of 64 x 64 PE's [3]. Goodyear Aerospace is currently constructing the massively parallel processor (MPP) for NASA [4]. The MPP consists of a matrix of 128 x 128 PE's, with 8 on a LSI chip.

The usual approach is to set up systems using the largest number of processors compatible with technological and economic constaints operating in parallel, by simplifying the single processing unit. Usually in these cases the arithmetic is bit-serial and each PE has few words of local memory. Up to now the maximum array built was the MPP, and also in this case, only a small portion of the practical image can be processed simultaneously, so the approach is to distribute the image to the processors one block a time. When performing local operations difficulties arise in the border processors, and in the case of iterative local operations, the useful part of the image propagates inward at every operation. In the future, VLSI technology will permit larger arrays, but it is unlikely that systems with a full image will be built in the next decade [5]. The impact of VLSI technology on these machines is limited by the fact that the number of processing elements integrated in one chip is limited by the number of pins required for the connection to all neighbors.

An alternative solution, utilized in diff3 and PHP II machines [5], has a number of powerful PE's, each one having a parallel neighborhood logic that provides access to the bits of its neighbors. Due to the moderate number of PE's, a suitable approach is that of distributing the PE's throughout the image, so that each one has a portion of the image. Obviously, an overhead due to the border problem and to the coordination of the PE's must be taken into account also in this case. During the 80's the complete experimentation and consolidation of ultra high speed technologies, such as the subnanosecond gallium arsenide, will possibly provide for this machine performaces comparable to those of the previous approach [5].

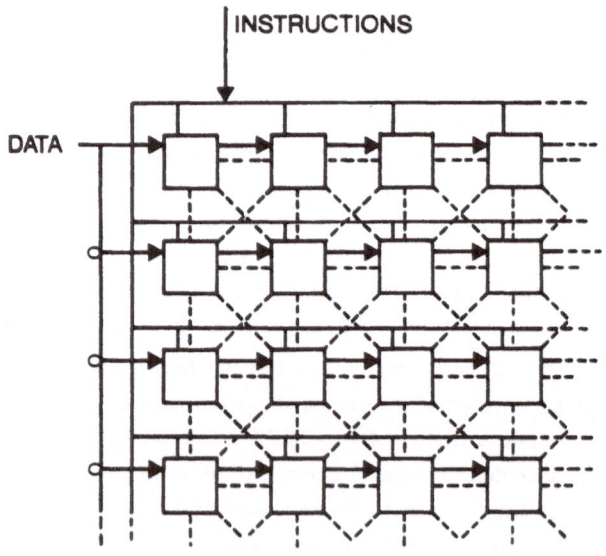

FIG. 2 - Interconnection in an SIMD array with square plane tesselletion and eight connected neighbors (CLIP IV).

The efficiency of SIMD architectures is drastically reduced when communication between PE's cannot be realized in a "tightly orchestred way". In [6] it has been shown that in several practical cases the SIMD structure has a communication time higher than its computation time. Methods for efficient propagation across cellular array have been developed and one of the most interesting is that of using a "pyramid" structure as in PCLIP designed by Tanimoto [7], with each plane of PE's half the size of the preceding layer Each PE is connected to a PE of the "father" plane and to four PE's of the "son" plane. In this case the speed of communication is increased at the expense of an increment of connection complexity and a higher number of PE's (30% more than in the array structure). However a higher computational speed can be obtained, due to the pyramid structure, in several practical applications [8].

3. PIPELINE ARCHITECTURES

Highly parallel systems with less flexibility then the cellular array have also been developed using pipeline techniques. An example of this type is the Cytocomputer designed by Sternberg [9] which in the prototype contains over 100 stages and in the more recent LSI version, with one stage on each chip, could involve over 500 stages (see Fig 3). The Cytocomputer is currently being exploited by ERIM. Also of this type is the systolic processor which is currently being developed by ESL [10].

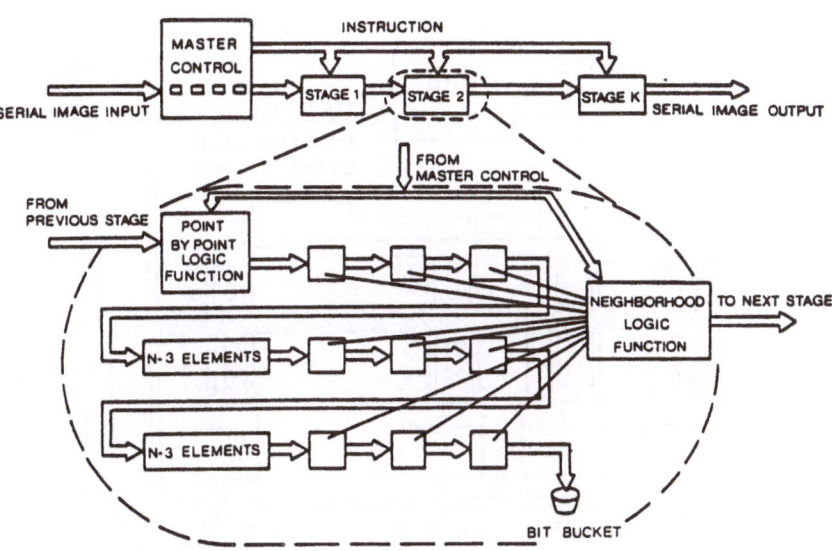

FIG. 3 - Example of a pipeline architecture: the Cytocomputer.

In pipeline architectures data streams in raster scan format from the buffer memory into a pipeline of processing stages. A sequence of delay units provides the access to defined subarrays of data (usually 3 by 3 square). Ignoring an initial delay in which the first pixel has traversed all the processors, as data streams in, processed data streams out: the total computation time will depend linearly on the number of pixels and not on the computational depth.

The subarray units are linked together in a linear unilateral structure, and the computation is decomposed a priori in elementary subtasks performed respectively in a sequential mode as the data flows downstream. If there are fewer program steps than processors, a number of "idle" elements pass the data they receive to the output and the machine is not fully utilized. On the contrary, if there are not enough units available, data has to be recycled with remaining steps loaded. In any case, for dedicated image processing systems where the processing is well defined and understood, this architecture offers the greatest economy and speed.

4. BUS ORIENTED ARCHITECTURES

A common solution for data transfer in a distrubuted system is provided by the bus structure. On the bus a logical link creates a communication path between modules. Buses represent shared resources and in tasks with frequent data transfer contention problems may arise (obviously communication proceed one at a time).

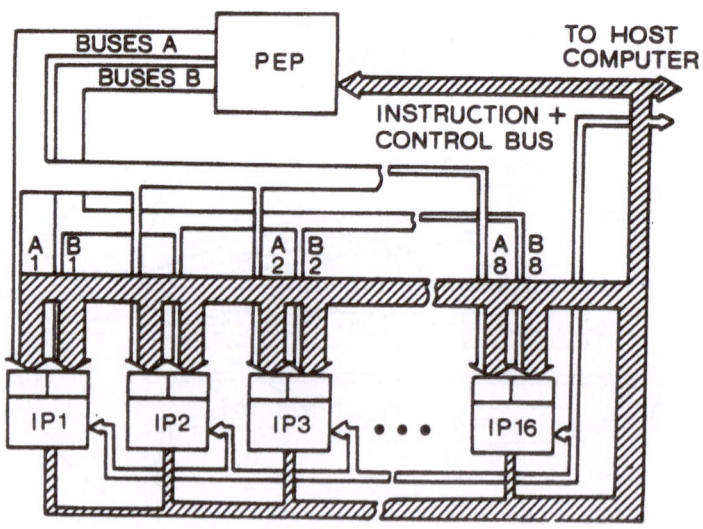

FIG. 4 - The FLIP system architecture.

Bus oriented architecture may be broadly subdivided in two classes: the former referring to architectures which contain homogeneous programmable modules, and the latter referring to architecture composed by heterogeneous or functionally dedicated modules.

Two examples of architectures of the former class are FLIP and ZMOB. In FLIP, designed by Gemmar [11], there are 16 homogeneous modules. However, if the number of processors is also moderate, the interconnection network is non-trivial . Each processor has two input data streams of 16x8 bits and one 8 bit-data output, that can be delivered to all other processors. Contention problems cannot arise but a very high number of buses is required (see Fig. 4). The other important system is ZMOB under development at the University of Maryland, a project initiated by Rieger et al. [13]. This system is intended for artificial intelligence applications including low level image processing. It consists of 256 Z80 microprocessors (64 Kbytes of local memory) connected in a synchronously 48 bit wide "conveyor belt" ring iwith a 10 MHz clock to a host VAX-11 computer.

Referring to the previous classes of SIMD architectures, in these cases the degree of concurrency in communication is lower, but higher speed buses with larger bandwidth are possible, so that the global computation times are often comparable.

The machines of the latter group consist of a host system and a set of special function processing units (SFU). The host system contains a host computer, a high speed image memory system and a high speed data bus. An SFU is a special purpose hardware for implementing a single function or a set of related functions. Each SFU has access to the high speed data bus, and may have considerable local memory in order to reduce the load on the data bus. In these machines communication is mainly concerned with the transfer of data to and from the units, and synchronization is necessary only for initiating and terminating functions.

Two examples of these systems are TOSPICS (see Fig. 5), Mori et al. [13] and PICAP II (see Fig. 6), Kruse et al. [14]. PICAP II is a multiuser system which contains an high speed asynchronous 32 bit data bus (40 Mbytes/s). The image memory consists of up to 16 256 Kbyte modules and the functional units include video and display processors, logical neighborhood and segmentation processor modules and a filter processor (FIP) composed of four 8-bit PE's which operate in SIMD mode and contain three pipeline stages each. TOSPICS is an interactive system built around a TOSBAC-40c minicomputer. The image memory is organized in 512 x 512 frames of 4 x 8 bit per pixel, and four graphic planes. A data bus of 4 Mbytes/s supports the access of the parallel pattern processor (PPP). The functional units of the PPP are: an address generator (a random access input stream and a random access output stream are supported concurrently), a 2D convolver capable of implementing an 8 x 8 convolution in 8 cycles, a region labeling module, a logical filtering module composed by the 3 x 3 neighborhood subarray, and a 256 x 16 table memory for look up table operations.

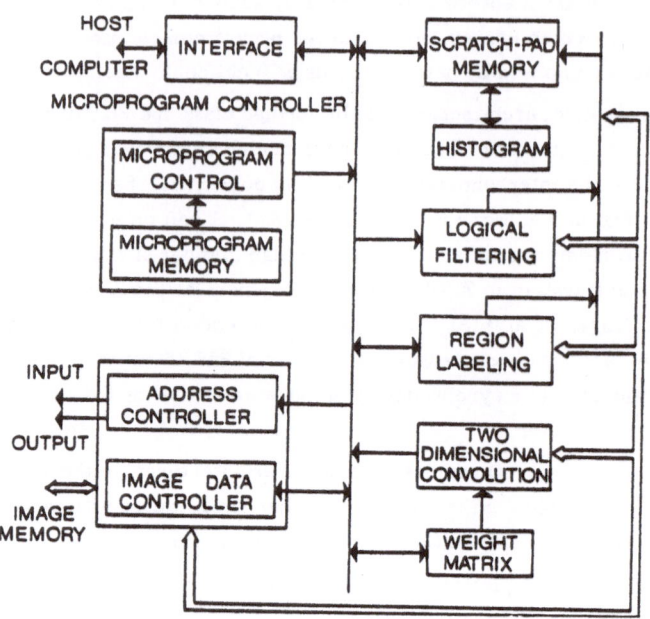

FIG. 5 - The PPP system architecture.

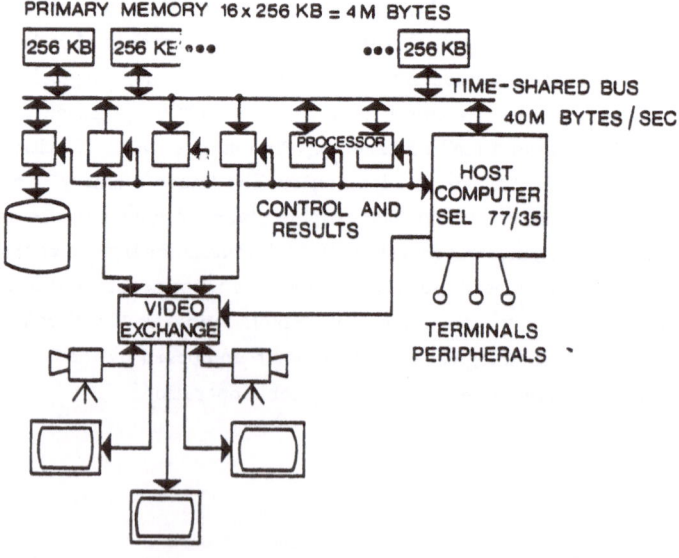

FIG. 6 - The PICAP II system architecture.

5. ARCHITECTURES WITH RECONFIGURABLE INTERCONNECTION STRUCTURES

Several computer architecture research groups are now turning their attention to the total image analysis problem. One of the main problems here is that image analysis is much less well understood than low level image processing and there is not a well defined set of general purpose algohrithms on which to base the design of an architecture. An approach taken by some research groups is to combine the concepts of SIMD and MIMD systems. Essentially, this means having an MIMD system with a selectable synchronizing mechanism to achieve the rapid data interchange characteristic of an SIMD system.

In this connection a number of reconfigurable interconnection structures have been proposed such as: the crossbar network (N^2switching element with a delay equal to one level of switching), the Delta, the Omega and the Banyan networks ($0.5Nlg_2N$ switching element with a delay equal to lg_2N levels of switching), etc. Systems of this class can be reconfigured into different forms of SIMD, MSIMD or MIMD machines. Obviously, the possibility of different modes of operation enables the system to "match" computing or data flow structures. Nevertheless, these architecture are not very efficiently organized for low level image processing, and the common neighborhood acces requested for every pixel in this kind of problem can become a real bottleneck.

A design of the reconfigurable SIMD/MIMD type is PASM which has been developed by Siegel et. al. [15] at Purdue University. PASM is conceived to consist of 1024 processing elements organized in 16 groups; each group has its own control unit.

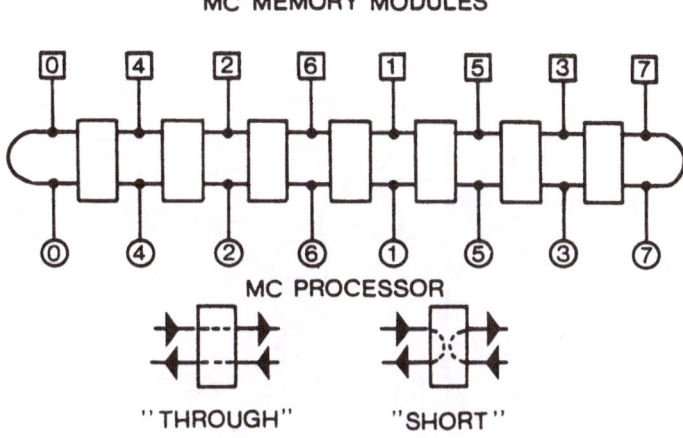

FIG. 7 - A reconfigurable bus scheme for interconnecting microcontroller processors and memory modules, PASM like.

A group may be considered to be a single SIMD processor with 64 PE's. Adjacent groups may be dynamically configured to behave as a single SIMD system. This system has been designed for general image processing applications. Presumably for low level image processing tasks the system would be configured as a single SIMD system of 1024 PE's. Much of the design effort has gone into the interconnection network which is of the permutation type (see Fig. 7).

A more recent design is the Array-Net which is currently being constructed at the University of Wisconsin-Madison by Uhr et.al[16]. This system consists of 256 PE's organized in groups of 16; each group having its own control unit. Each group is organized as a 4x4 matrix of PE's and the groups themselves are also organized as 4x4 matrices. When all control units have the same program contents and are synchronized then the system behaves like a single SIMD system having a 16x16 matrix of near-neighbor connected PE's. A single PE has 8-bit data-paths; however, a whole group can be reconfigured to behave as a single processor, combining several PE's to form a wider datapath. In this mode the system can operate as an MIMD system with 16 independent processors, combining several PE's to form a wider datapath. In this mode the system can operate as an MIMD system with 16 independent processors; each processor can execute a separate program.

Other parallel systems which are relevant to high level processing are large scale MIMD systems. Early work in this area was done at Carnegie-Mellon University with C.mmp based on PDP-11 computers [17] and CM* based on LSI-11 computers [18]. An important system of this type is MICRONET developed by Wittie [19]. This is a system of microprocessors (initially LSI-11's) which is designed to investigate problems in MIMD communication. It is not directed to any particular application but up to now it is one of the very few such systems to be constructed.

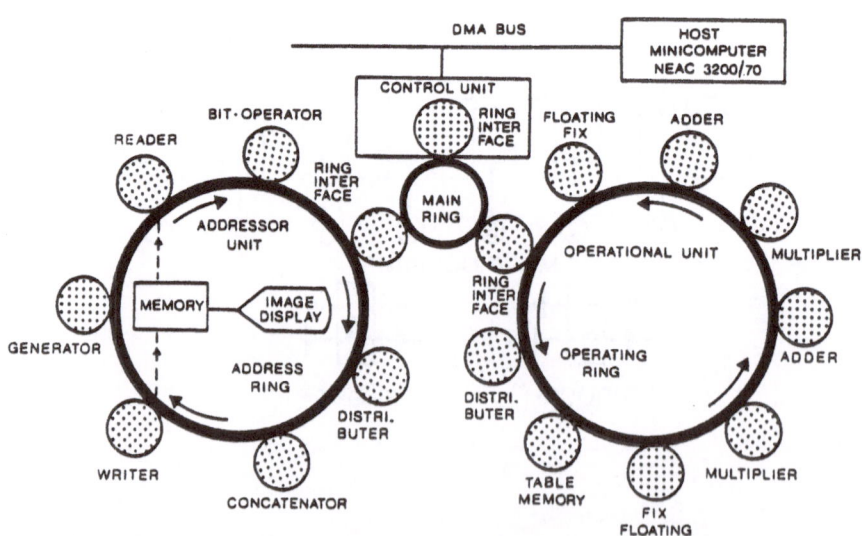

FIG. 8 - Block diagram of the TIP system.

6. FUTURE SYSTEMS

Among the recently introduced systems it is worthwhile to point out two new machines: TIP by Hanaki et al. [20], and PUMPS by Briggs et al. [21]. The Template-controlled Image Processor is a data flow machine (see Fig. 8), especially designed for image processing applications.

TIP is essentially composed of three rings: the operational one containing special function units, the addressor one, and the main ring which interconnects with the other two. Data flows through the ring, each data item has one identifier and one or more destination flags. When the data reaches the right functional unit an operation is performed according to the template the data identifier matches. This seems to be a new interesting architecture. In the near future we will see if it is appropriate for image processing purposes.

The PUMPS is a system designed for general image processing and pictorial database management (see Fig. 9). PUMPS is composed of: a set of MIMD/SISD Task Processor Units (TPU) with local memories; a shared memory connected to TPU's by a delta network ($0.5Nlg_2N$ switching elements with a delay equal to log_2N levels of switching), a set of special function units implemented with VLSI modules, a crossbar interconnection network between TPU's and special function units to implement a macropipeline (each stage can be a SISD, MIMD, SIMD, pipeline or SFU). Each TPU can perform multitasking and has local caches, so communication between TPU's and shared memory is in data block mode.

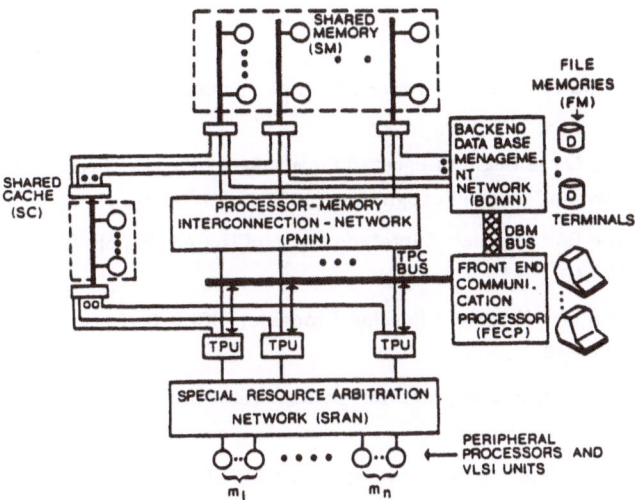

FIG. 9 - The PUMPS system architecture.

This is the first system that has been designed for very different pplications. Its architecture is compatible with both low level and high level image processing tasks. Furthermore, it is one of the first systems that includes pictorial data base management. PUMPS is currently restricted to the design and simulation levels, but so far the hardware implementation has not been initiated.

7. CONCLUSIONS

Since the 1960s a wide variety of computer architectures for image processing and pattern recognition have been designed and in many cases built. In this paper a selection of systems has been introduced focusing attention on different structural characteristics and on the impact that image processing and pattern recognition tasks have on these architectures.

Summarizing we can say that to a large extent, the computer architecture problem for low level image processing has now been solved, and implementation is in the realm of private industry. The MIMD systems, conceived for high level image processing are instead research projects at an early stage of developement that could however result in some very powerful systems in the not too distant future.

REFERENCES

1. S. H. Unger, "A Computer Oriented Toward Spatial Problems" Proc. of the IRE, October (1958), pp 1744-1750.

2. M. J. B. Duff, "CLIP4: A Large Scale Integrated Circuit Array Parallel Processor" 3rd International Joint Conference on Pattern Recognition, (1976), pp 728-732.

3 S. F. Readdaway, "The DAP Approach" Infotech State of the Art Report on Supercomputers, Vol. 2, (1979), pp 836-840.

4. K. E. Batcher, "Design of a Massively Parallel Processor" IEEE Trans. on Computers Vol. C-29, No 9 (1980), pp 836-840.

5. K. Preston, Jr. "Cellular logic computer for pattern recognition" Computer, Vol. 16, No 1 (1983), pp 36-47.

6. V. Cantoni, S. Levialdi, C. Guerra, "Towards an evaluation of an image processing system" in Computational structures for image processing, M. J. B. Duff ed., Academic Press, (1983), pp 43-56.

7. S. Tanimoto, "Towards a hierarchical cellular logic: design considerations for pyramid machines" TR-81-02-01, University of Washington, Seattle, (1081).

8. B. H. Mc Cormick, E. W. Kent, C. R. Dyer, " HIghly parallel structures for real time image processing" ISL-TR-VRL=13, University of Illinois at Chicago Circle, (1980)

9. S. R. Sternberg, "Parallel Architectures for Image Processing," Proceedings of the 3rd International IEEE COMPSAC, Chicago, (1979), pp 712-717.

10. D. W. L. Yen and A. V. Kulkarni, "The ESL Systolic Processor for Signal and Image Processing," IEEE Computer Society Workshop on Computer Architecture for Pattern Analaysis and Image Database Management, Hot Springs, Virginia, November 11-13, (1981), pp 265-272.

11. K. Luetjen, P. Gemmar, H. Ischen, "FLIP: a flexible multiprocessor system for image processing" Proc. 5th Int. Conf. Pattern Recognition, (1980), Miami, pp 326-328.

12. C. Rieger, "ZMOB: Doing it in Parallel," IEEE Computer Society Workshop on Computer Architecture for Pattern Analysis and Image Database Management, Hot Springs, Virginia, November 11-13, (1981), pp 119-214.

13. K. I. Mori, M. Kidode, H. Shinoda, H. Asada, "Design of local parallel processor fo IP" Proc. AFIPS Conf., Vol. 47, (1978), pp 1025-1032.

14. B. Kruse, B. Gudmundsson, D. Antonsson, "FIP: the PICAP II filter processor" Proc. 5th Int. Conf. Pattern Recognition, (1980), Miami, pp 484-488.

15. H. J. Siegel, et. al., "PASM: A Partitionable SIMD/MIMD System for Image Processing and Pattern Recognition," IEEE Trans. on Computers, Vol. C-30, No 12, December (1981).

16. L. Uhr, M: Thompson and J. Lockey, "A 2-Layered SIMD/MIMD Parallel Pyramidal Array/Net," IEEE Computer Society Workshop on Computer Architecture for Pattern Analaysis and Image Database Management, Hot Springs, Virginia, November 11-13, (1981), pp 209-216.

17. W. Wulf and R. Levin, "A Local Network," DATAMATION, Feb. (1975), pp 47-50.

18. R. J. Swan et al., "Cm* — A modular multi-microprocessor," AFIPS Conference Proceedings, Vol. 46, 1977 NCC, pp 637-644.

19. L. D. Wittie, R. S. Curtis and A. J. Frank "MICRONET/MICROS — A Network Computer System for Distributed Applications", in "Multicomputers and Image Processing: Algorithms and Programs", K. Preston and L. Uhr eds. Academic Press (1982) pp 307-318.

20. S. Hanaki, T. Temma, "Template-controlled Image Processor" in Multicomputer an Image Processing, K. Preston, L. Uhr eds., Academic Press, (1982).

21. F. A. Briggs, K. Hwang, K. S. Fu, M. Dubois, "PUMPS architecture for pattern analysis and image database management" Proc. Pattern Recognition and Image Processing Conf., Dallas, (1981), pp. 178-187.

MEMORY-COUPLED PROCESSOR ARRAYS FOR A BROAD
SPECTRUM OF APPLICATIONS

Gerhard Fritsch

Universität Erlangen-Nürnberg
Institut für Mathematische Maschinen
und Datenverarbeitung (III)
Martensstr. 3
D - 8520 Erlangen / F.R. Germany

Abstract:

An efficient use of a multiprocessor system requires appropriate
mapping of the problem structure onto the multiprocessor structure.
Two memory-coupled multiprocessor systems are presented and results
obtained from computation of a number of applications are reported.

1. INTRODUCTION

The user's demand for much higher computational power than nowadays
available will not be satisfied in the future by still more powerful
computers of the von-Neumann type, because of technological and physical limits. In order to achieve higher computational power an alternative is offered by parallel processing and parallel storing. Various
forms of parallel organization have been realized in modern computers.
Advances in very large-scale integration (VLSI) technology favor the
design of large parallel computers whose processing-storing-elements
are connected.

A rough classification of multiple processor organizations can be based
upon the degree of coupling between the processor-memory-modules. A
"coupling constant" T_{wc} can be defined as the "worst case processor's
minimum access time to a global data structure in the system"
[FULLER 78]. Thus, multiple processor organization can be grouped in

computer networks (T_{wc} range from 1 to 10^{-4} s), multiprocessors (10^{-4} to 10^{-7} s) and multi-arithmetic-logic-units (10^{-6} to 10^{-9} s).

The generalized term "multiprocessor" comprehends a rather large class of processor-memory-structures which have a high degree of resource sharing including sharing of all the primary directly addressable memory. This class can be subdivided with respect to the type of the interconnection network which connects the processor- and memory-modules, as for example crossbar switches, common busses or multiport systems. Examples for the latter are dealt with in this paper.

The inherent parallelism of a computer system can be characterized by the Erlangen Classification Scheme which considers three processing levels: program control units, arithmetic and logical units, elementary logic circuits [HÄNDLER 75, 77a, 77b].

Among the problem classes with nearly unlimited demand for computational power are condensed matter physics, plasma physics, quantum chemistry, astrophysics, nuclear physics, high energy physics, hydro-dynamics, aerodynamics, pattern recognition, geophysics, complex technical systems etc. Such ample variety of applications can mostly be reduced to linear algebraic problems or non-linear optimization problems. In general, both types of algorithms can be parallelized so that computation on multiprocessor systems, in particular on arrays of tightly coupled processor-memory-modules, can be achieved.

2. MUTUAL MAPPING OF PROBLEM AND COMPUTER STRUCTURES

With the advent of multiprocessor systems and parallel processing a variety of hardware and software structures have been proposed. Efficient computation requires appropriate mapping of the task structure of the user problem onto the multiprocessor structure or vice versa. Thereby, various requirements have to be met concerning the distribution of code and data over the processor-memory-modules in order to assume minimum interprocessor communication, minimum transport of global data etc. The development of modern computers, in particular multiprocessor systems, is influenced by interactions among hardware technology, computer architecture, software systems and applications. On one hand the complexity of the problems that can be solved, depends on the computer speed, on the other hand, new algorithms, in particular for parallel computation, may have an important effect on the design of new parallel computer architectures (Fig. 1).

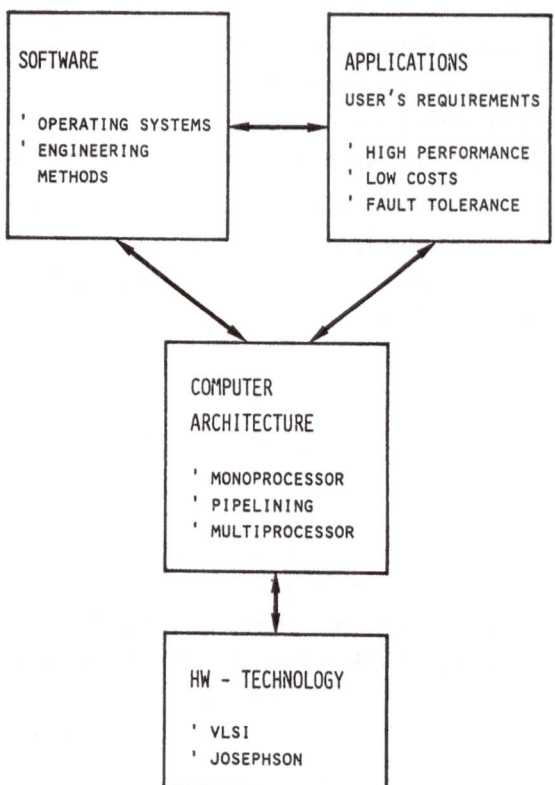

Fig. 1: Interactions between hardware technology, computer architecture, software development and user's demands.

As VLSI technology advances, increasingly powerful microprocessors and memories can be used as computer components, so that methods to inter-connect them efficiently will become important. In contrast to the classical monoprocessors, mutual mapping of problem and computer structures will become an important activity for the programmer of computers with new parallel architectures.

A larger class of computationally most intensive problems in natural and engineering science is the simulation of continuous fields. The

starting point is a mathematical model of, say, a physical phenomenon.
A related mathematical model may be represented by a set of partial
differential equations. In a next step a discretization method is
applied, i.e. a finite-difference method [RICHTMYER 67], a finite-
element method [STRANG 73] or a particle-mesh method [HOCKNEY 81].
Continuous space and time variables are replaced by a mesh and a sequence of discrete time steps respectively, and continuous physical
variables are replaced by arrays of values. The set of differential
equations is transformed into a set of algebraic equations. By this
algebraic approximation, in each iteration new values of the physical
variables are computed out of the old values in adjacent neighbors of
each point. Since the operations for different points are independent,
they can be carried out concurrently. An array of coupled processors
is well suited to compute this inherently parallel algebraic problem.
Each processing module can be associated with a subregion of the mesh
or - in special cases - with a mesh-point. Furthermore, with tightly
coupled processor arrays, data exchange between neighboring processors
which are associated with neighboring subregions of the mesh can easily
be achieved. Fig. 2 illustrates the mutual mapping between 2- and 3-
dimensional grid-like user problems and processor arrays. In the following two chapters two multiprocessor systems and the results of the
implementation of user problems are presented.

3. EGPA - ERLANGEN GENERAL PURPOSE ARRAY

3.1 Principles of architecture and operation

The architecture of the EGPA-multiprocessor system was proposed in
1975 [HÄNDLER 75]. The essential features and design objectives of
this general purpose architecture are summarized in the following:

(1) The architecture consists of processor-memory-modules (PMM) which
    are connected in two-dimensional orthogonal grid-like structures,
    thus forming a "plane". Two or more planes are arranged hierar-
    chically such as to represent a "pyramid".

(2) In order to achieve higher computing power the multiprocessor system
    can be freely extended so that the global complexity of the inter-
    connection system increases only linearly with the number of PMMs

MAPPING OF STRUCTURES

USER PROBLEM   PROCESSOR ARRAY

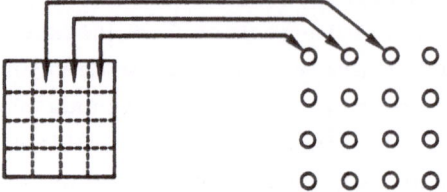

i-th TIME STEP
Parallel computing
over the defined area
(i+1)-th TIME STEP

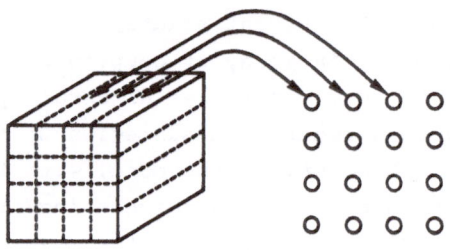

i-th TIME STEP
Simultaneous computa-
tion of one layer (as
in the 2-dim. case),
then next layer and so
on ...
(i+1)-th TIME STEP

Fig. 2: Computing grid-like user problems by regular processor-memory
        arrays. Application to 2- and 3-dimensional problems from con-
        densed matter physics, hydrodynamics, aerodynamics, etc.

conserving constant local complexity.

(3) Each processor has access to the memories of the four adjacent
    PMMs in the same plane (bi-directional connections) and to the me-
    mories of four subordinated PMMs (unidirectional connections).
    Therefore, each PMM - except those at the top level - can be acces-
    sed by a supervisor processor.

(4) The structure of the operating system corresponds to the hierarchi-
    cal hardware structure. At the lowest level - the worker processor
    array - user problems are processed. At higher levels administra-
    tive functions, including I/O, are performed.

(5) The multiprocessor system can operate in three modes:

- Independent Mode: The system is partitioned into separately operating computers.
- Concurrent Mode: A task is subdivided into subtasks which are processed separately by different PMMs. Coordination is assured by supervising PMMs.
- Dataflow Mode: Operation by functionally oriented data flow, e.g. macro-pipelining [HÄNDLER 73].

With each mode both "horizontal" (or conventional) processing and "vertical" processing can be achieved. The idea of "vertical" processing, i.e. a pseudo-associative processing, is based on an unconventional interpretation of the bit position in the main memory [HÄNDLER 73; BODE 80, 83]. The conventionally "horizontally" stored data of wordlength w are considered as "bit-slices" in a fixed bit position of w "vertically" stored data of arbitrary wordlength.

3.2 The Pilot-Pyramid

Fig. 3 depicts the above explained interconnection system for the realized experimental EGPA-pilot pyramid: Mutual memory access between neighboring PMMs at the same level (A-processors) and unidirectional access of the supervisor PMM (B-processor) to the memories of the A-processors. The PMMs consist of commercially available computers AEG 80-60 (32 bit wordlength, microprogrammable, multiport memory). The operating system is a hierarchical multiprocessor operating system based on the original uniprocessor operating system. Interprocessor communication takes place via common control blocks and mailbox-techniques. In order to improve processor-processor-communication, additional connections between all processors allow for interprocessor-communication by interrupts.

Parallel processing was tested with a number of quite different applications. Parallel algorithms were developed and their subtask structure was mapped onto the EGPA-system. Various problems were implemented and the speedup - versus a monoprocessor - was measured. The results are summarized in the TABLE. The limiting speedup of the EGPA-system is four, as this system contains four worker processors. In order to be able to improve parallel computations, the application of evaluation

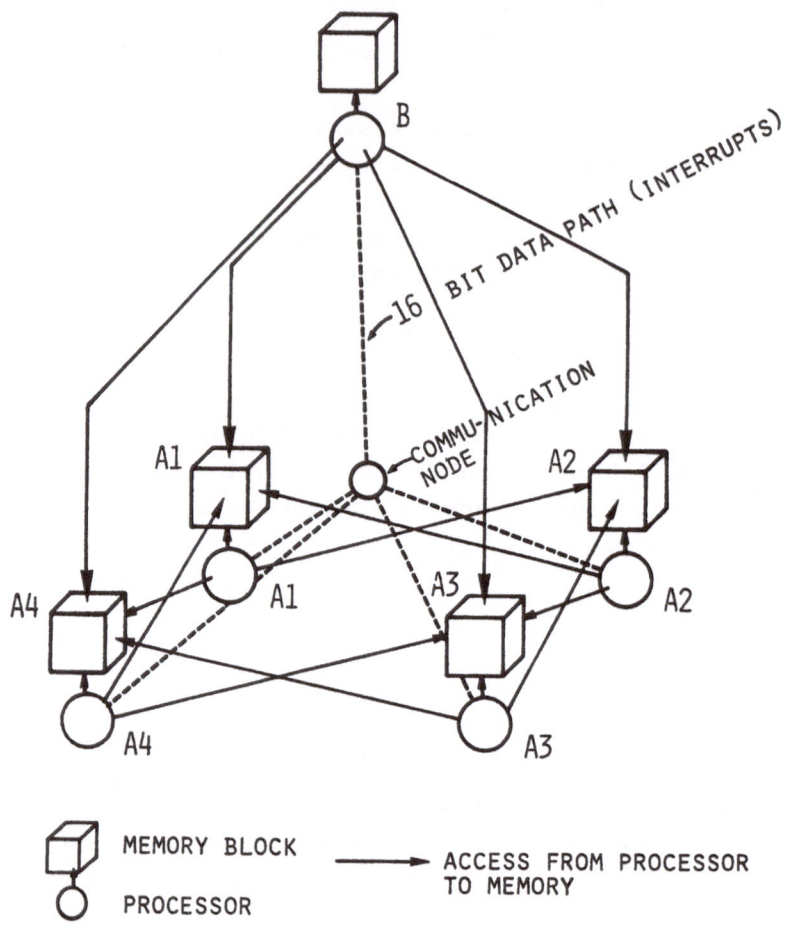

Fig. 3: Pilot pyramid of the EGPA-multiprocessor system:
Each node consists of one processor (circle) and one memory
block (square).

Interprocessor communication:
(1) via common memory (mailbox technique),
(2) processor-processor-interrupt coupling (via communication
node)

methods is required. Hardware and software evaluation tools were
developed and integrated in the EGPA pyramid [FROMM 83].

Linear algebra [HENNING 83]   SPEED UP
- Matrix inversion (2oo x 2oo dense)
  Gauss - Jordan   3.8
  column-substitution   3.9
- Matrix multiplication (2oo x 2oo)   3.7
- Solving of linear equations
  Gauss-Seidel   ca. 4.o

Differential equations [FROMM 82]
- Relaxation   ca. 3.5

Image processing and graphics
- Topographical representation [KNEISSL 82]   3.6
- Illumination of the topographical model   2.4
- Line following   ca. 2.9
  (vectorizing of a grey level matrix)
- Distance transformation [GOESSMANN 83]
  each processor is working on a fixed
  part of data   ca. 3.o
  dynamic assignment of varying parts of
  data   ca. 3.3

Non linear programming [FRITSCH 81]
- Search for minima of a multi-dimensional
  object function given by an algebraic
  term   ca. 3.2

Graph theory
- network flow with neighborhood support   3.5
  (each idle processor helps one of its
  neighbors)

Text formating [RATHKE 83]   2.6

Maximal theoretical speedup (4 array processors) 4.o

TABLE: Applications implemented on the EGPA-multiprocessor
       system

Linear algebra [HENNING 83]   SPEED UP
- Matrix inversion (2oo x 2oo dense)
  Gauss - Jordan   3.8
  column-substitution   3.9
- Matrix multiplication (2oo x 2oo)   3.7
- Solving of linear equations
  Gauss-Seidel   ca. 4.o

Differential equations [FROMM 82]
- Relaxation   ca. 3.5

Image processing and graphics
- Topographical representation [KNEISSL 82]   3.6
- Illumination of the topographical model   2.4
- Line following   ca. 2.9
  (vectorizing of a grey level matrix)
- Distance transformation [GOESSMANN 83]
  each processor is working on a fixed
  part of data   ca. 3.o
  dynamic assignment of varying parts of
  data   ca. 3.3

Non linear programming [FRITSCH 81]
- Search for minima of a multi-dimensional
  object function given by an algebraic
  term   ca. 3.2

Graph theory
- network flow with neighborhood support   3.5
  (each idle processor helps one of its
  neighbors)

Text formating [RATHKE 83]   2.6

Maximal theoretical speedup (4 array processors)   4.o

TABLE: Applications implemented on the EGPA-multiprocessor
       system

3.3 Parallel computation of an optimization problem

As an example, the parallel computation of a minimum search problem on the EGPA system will be discussed in this section. The user problem consists of the calculation of the reaction and transport parameters of a chemical reaction system. To that end, the global minimum of an objective function has to be determined. It is a least-squares function containing the sum of the squares of the differences between the experimental and the theoretical values of temperature and concentration.

For minimization of the objective function the simplex-method of Nelder and Mead [NELDER 65] was applied. This minimum search procedure is a stepping method: A simplex moves in the function landscape changing its shape by replacing the vertix with the highest function value by a new one with a lower function value. In this way the simplex proceeds towards the function minimum. The original procedure is sequential. It can be parallelized in various ways, e.g. by searching minima on different routes simultaneously. However, parallelized versions can prove less efficient than the sequential one. The number of function evaluations necessary to find a given minimum can be taken as a standard of comparison. In case the sequential minimum search algorithm is expected to work more efficiently than the parallel one, another, quite trivial, parallelization strategy can be applied. The minimum search area is subdivided into as many subareas as worker processors are available. Each processor is allocated to one subarea and the sequential minimum search algorithm is applied. Dependent upon the function landscape and the halting criterion the number of function evaluations can vary considerably from one subarea to the other. This can produce an unbalance of the computing load between the processors. The unbalanced load can be smoothed out by applying neighborhood-aid, that is using idle neighboring worker processors to support still busy neighbors. This processor allocation strategy can be efficiently applied only with tightly coupled processor arrays. An example is given in Fig. 4.

Subdivided
minimum search area

EGPA
worker processor array

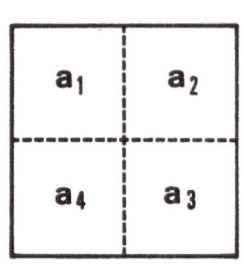

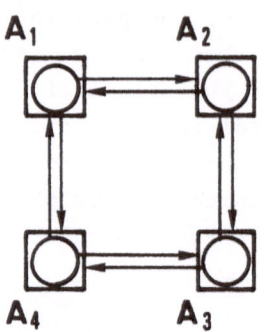

ALLOCATION OF THE PROCESSORS
TO THE SUBAREAS (SUBTASKS)

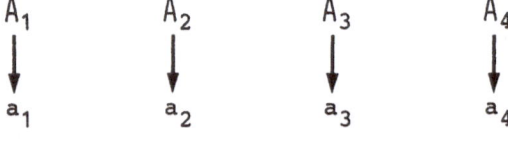

	A_1	A_2	A_3	A_4	
	↓	↓	↓	↓	
	a_1	a_2	a_3	a_4	
EXAMPLE:	589	354	402	1503	FUNCTION EVALUATIONS

WITH NEIGBORHOOD AID

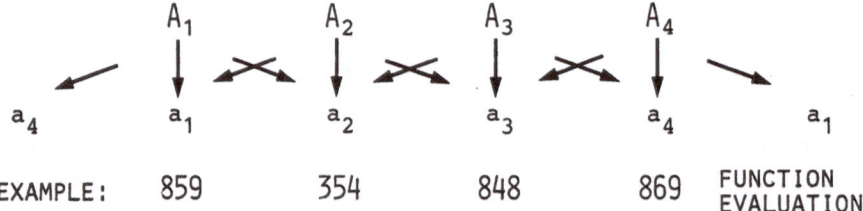

		A_1		A_2		A_3		A_4	
a_4		a_1		a_2		a_3		a_4	a_1
EXAMPLE:		859		354		848		869	FUNCTION EVALUATIONS

Fig. 4: Allocation of the worker processors to the subtasks of the
optimization problem.
Special case: Subdividing the minimum search area and mapping
the subareas onto the processor array, without and with neighborhood aid.

3.4 An extended EGPA-system

The EGPA-architecture allows for the extension towards more powerful multiprocessor systems by adding one or more larger processor arrays at the bottom of the pyramid. At the Computer Science Department (IMMD) of the University of Erlangen-Nürnberg the Erlangen Multiprocessor System 85 (EMSY 85) has been conceived on the base of the experiences gained with the EGPA-project. Four arrays of 1, 4, 16 and 64 PMMs respectively will be arranged hierarchically. The elementary structure is a pyramid equivalent to the EGPA-pilot pyramid. Each PMM will consist of an iAPX 286/287 microprocessor and a one-half-megabyte multiport memory. The operating system based on UNIX will be organized analogously to the hardware structure. The operating system is to be distributed over the EMSY 85-pyramid, increasing in power towards the top [FRITSCH 83].

The EMSY 85-pyramid is depicted in Fig. 5. Besides, a possible mapping strategy for a hierarchical program system onto EMSY 85 is represented. The program system refers to the minimization problem discussed in the preceding section. This is an example for parallel computation at two levels: The subroutine of the minimum search procedure runs on the 16-PMMs-array while the objective function is computed on the large worker processor array with 64 PMMs. The structure of the parallel program system is shown in Fig. 6. For the purpose of clearness of the representation the parallel program system is structured for computing on a multiprocessor system of the EGPA-type, consisting of 16 A-processors, 4 B-processors and 1 C-processor.

4. DIRMU - DISTRIBUTED RECONFIGURABLE MULTIPROCESSOR KIT

The aim of the DIRMU-project has been to offer a system kit of plug-in processor-memory-modules (PMMs) of only one type in order to configure user-definable special purpose multi-microprocessor systems tailored to a specific class of applications [HÄNDLER 80]. The building block of DIRMU-configurations consists of a processor submodule and a memory submodule. The processor submodule contains the microprocessor (Intel 8086/8087), some private ROM (local operating system, self-test programs), private RAM and I/O-functions. The memory submodule is organized

170

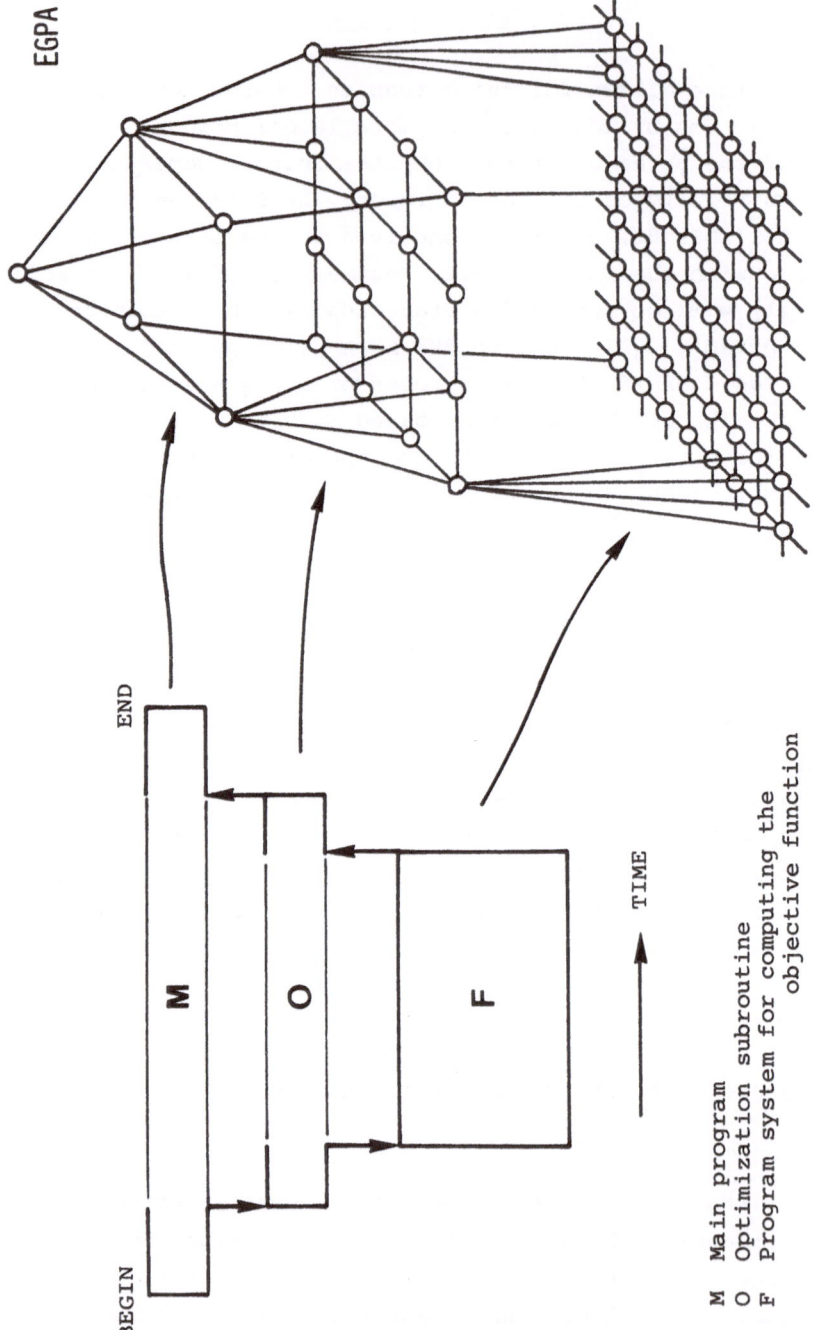

M   Main program
O   Optimization subroutine
F   Program system for computing the
    objective function

Fig. 5: Mapping a hierarchical program system onto a tightly coupled hierarchical multi-
        processor, EMSY 85 (Erlangen Multi Processor System 85), conceived on the base of
        the experiences gained with EGPA. Some of the elementary pyramids are highlighted.

PROCESSORS   Parallel Program

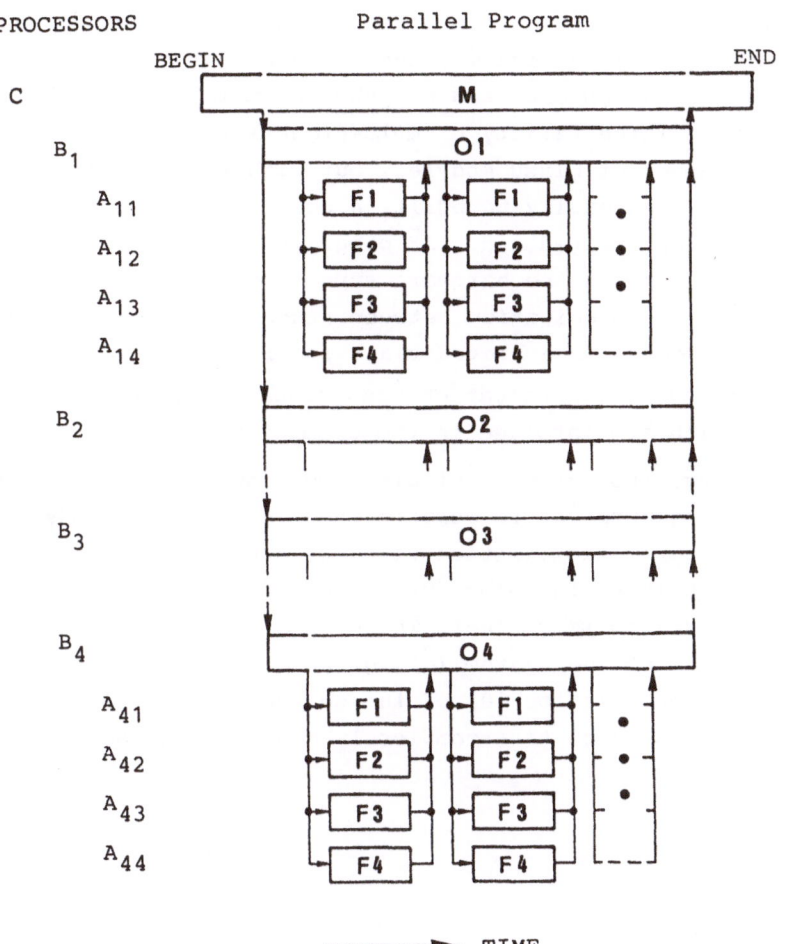

Fig. 6: Parallel computation of a minimum search problem at two
        levels:
        M main program,
        01, 02, 03, 04 concurrently running optimization subroutines,
        F1, F2, F3, F4 concurrently running subroutines for computing
                        the objective function
        The parallel program can be mapped onto a EGPA-type multipro-
        cessor system consisting of 16 A-processors, 4 B-processors,
        1 C-processor.

as a multiport memory, which can be accessed by its "own" processor submodule and by a set of other processor submodules. The connections between processor submodules and neighboring memory submodules are plugable. Thus a large variety of multiprocessor configurations with an arbitrary number of DIRMU modules, but with restricted neighborhood up to 7 modules, can easily be built. A commercially available operating system is implemented and test programs are used for fault recognition. Fig. 7 depicts the hardware structure of a DIRMU module and Fig. 8 represents communication between processes. Communication within a processor submodule is achieved via a local mailbox in the private memory. Inter-module communication uses global mailboxes in the memory-submodules, which are involved in the communication.

In the following, two examples of DIRMU-configurations are given. Fig. 9 presents a configuration to compute the minimum of an objective function. Structural similarities with the EGPA-configuration (section 3.3) are obvious. The other user problem deals with an automatic transportation system (Fig. 10a). This task can be subdivided into 3 subtasks: Control of the system (C), graphic representation of the system (G) and optimization (O) of the charge distribution. The three subtasks can be assigned to a completely connected 3-module configu-

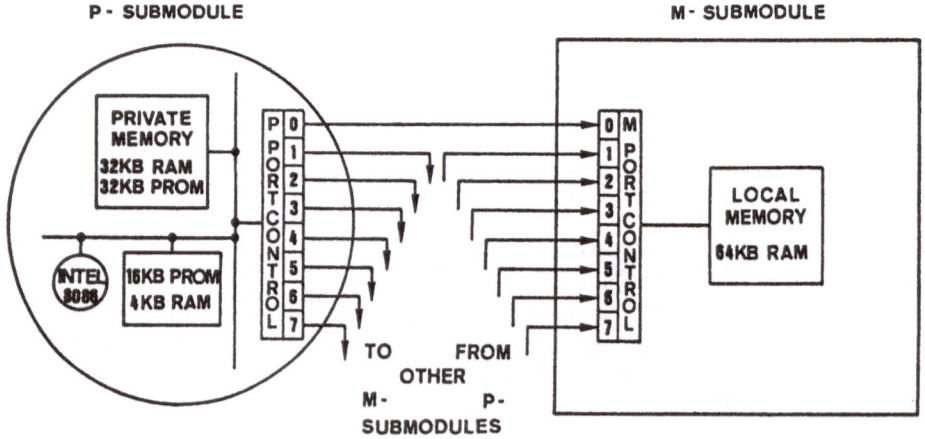

Fig. 7: DIRMU - Module: P(rocessor) - Submodule,
M(emory)   - Submodule,
P-Ports,
M-Ports

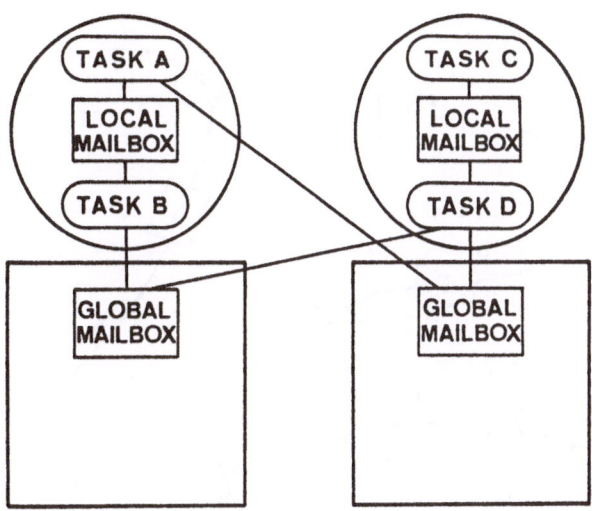

Fig. 8: Communication between processes in the DIRMU-System:
  (a) Within a processor submodule via a local mailbox
      (private memory).
  (b) Between two DIRMU-modules via global mailboxes.

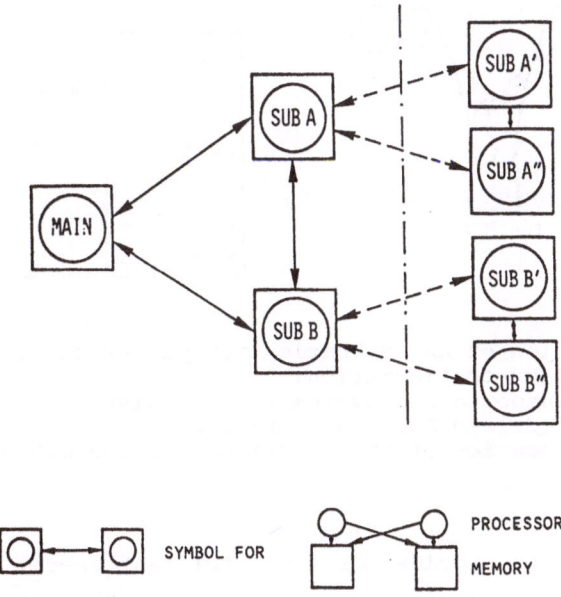

Fig. 9: DIRMU-Configuration for computing a minimization problem:
        Module "MAIN" for the main program, the optimization sub-
                       routine and I/O,
        Modules "SUB A" and "SUB B" for the computation of the
                       objective function,
        4 auxiliary modules supporting the computation of the objec-
        tive function (e.g.: integration-subroutines)

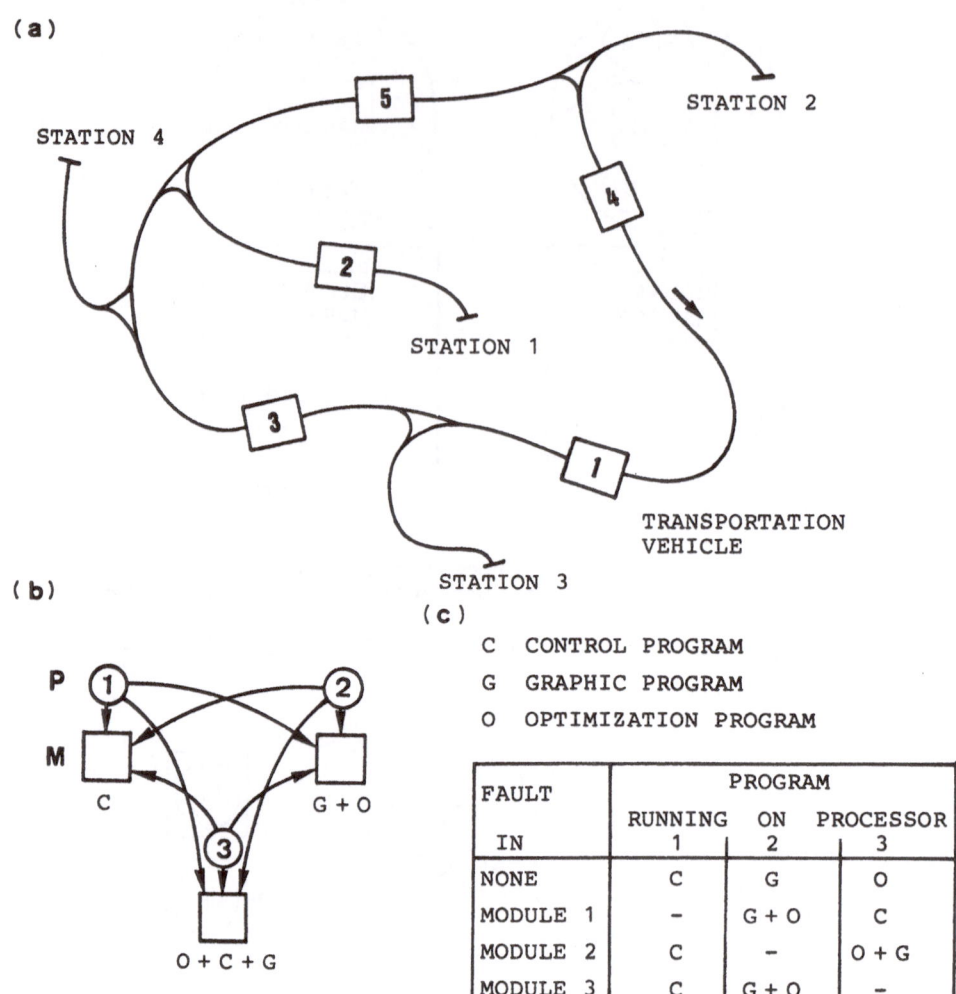

(a)

STATION 4

5

STATION 2

4

2

STATION 1

3

1

TRANSPORTATION
VEHICLE

STATION 3

(b)

(c)

C   CONTROL PROGRAM

G   GRAPHIC PROGRAM

O   OPTIMIZATION PROGRAM

FAULT	PROGRAM		
	RUNNING	ON	PROCESSOR
IN	1	2	3
NONE	C	G	O
MODULE 1	–	G + O	C
MODULE 2	C	–	O + G
MODULE 3	C	G + O	–

Fig. 10: Simulation of an automatic transportation system and control
by a DIRMU-configuration:
(a) Transportation system for the distribution of goods
(b) 3-Module-DIRMU-configuration
(c) Allocation of the programs onto the DIRMU-configuration

ration. In order to provide for fault-tolerant operation the programs
C, G, O have to be stored twice, in different PMMs (Fig. 10b). If a
fault is detected in one PMM, this defective module has to be removed

from the system and the subtasks must be redistributed to the operative PMMs, as for instance prognosed in Fig. 10c [HALLER 83].

5. CONCLUSIONS

The discussion about benefits and drawbacks of numerous architectural conceptions of multiprocessor systems has to take into account the user's demands. The goal is to achieve more computational power. An important criterion for the usefulness of a given multiprocessor architecture is whether the user problem can be decomposed in such a way that its subtask structure can be efficiently mapped onto the computer structure or vice versa.

Most applications with a field-like structure can be mapped onto tightly coupled processor arrays. The use of a powerful multiprocessor operating system requires a regular and hierarchical hardware architecture. Furthermore, such a multiprocessor system should be extensible in order to meet future demands for higher computational power. Two systems were presented, one with a regular, hierarchical architecture and another one with a flexible structure tailored to the needs of the application. Each system is built with one type of processor-memory-module.

6. REFERENCES

[BODE 80]   Bode, A: Vertical Processing: The emulation of asso-
               ciative and parallel behavior on conventional hardware,
               in Microprocessor Systems, EUROMICRO 80, North-Holland
               Publ. Comp. 1980

[BODE 83]   Bode, A, W. Händler: Rechnerarchitektur II,
               Springer Verlag 1983

[FRITSCH 81]   Fritsch, G., H. Müller: Parallelization of a minimisa-
               tion problem for multiprocessor systems, Lect. Notes
               in Computer Science, No. 111 (Ed. W. Händler), 453 -
               463, Springer-Verlag 1981

[FRITSCH 83]   Fritsch, G., W. Kleinöder, C.U. Linster, J. Volkert:
               EMSY 85 - The Erlangen multiprocessor system for a
               broad spectrum of applications, Proc. 1983 Int. Conf.
               Parallel Processing, IEEE Comp. Soc. Order No. 479 (Ed.
               H.J. Siegel and L. Siegel), 325-330, IEEE Computer
               Society Press 1983

[FROMM 82]   Fromm, H.J.: "Multiprozessor-Rechneranlagen: Programm-
                strukturen, Maschinenstrukturen und Zuordnungsprobleme",
                Arbeitsberichte des IMMD, Univ. Erlangen-Nuernberg,
                Band 15, Nr. 5, 1982

[FROMM 83]   Fromm, H.J., U. Hercksen, U. Herzog, K.H. John, R.
                Klar, W. Kleinöder: Experiences with performance
                measurement and modeling of a processor array, IEEE
                Trans. on Computers, Vol. C-32, No. 1, 15-31, 1983

[FULLER 78]   Fuller. S.H.. J.K. Ousterhout. L. Raskin. P.I. Rubin-
                feld, P.J. Sindhu, R.J. Swan: Multi-Microprocessors,
                an Overview and Working Example, Proc. IEEE, Vol. 66,
                No. 2, 216-226 (1978).

[GOESSMANN 83] Goessmann, M., J. Volkert und H. Zischler: "Image
                Processing and Graphics on EGPA", EGPA - Internal
                Paper (to be published)

[HÄNDLER 73]   Händler, W.: A concept of macro-pipelining with high
                availability, Elektron. Rechenanlagen, Vol. 15, 269-
                274 (1973)

[HÄNDLER 74]   Händler, W.: Unconventional computational equipment,
                Arbeitsberichte des IMMD, Universität Erlangen-Nürn-
                berg, Vol. 7, No. 2, 1974

[HÄNDLER 75a]   Händler, W., R. Klar: Fitting processors to the needs
                of a General Purpose Array (EGPA), Proc. Micro 8,
                Chicago, Sept. 21-23, 87-97 (1975)

[HÄNDLER 75b]   Händler, W.: On classification schemes for computer
                systems in the post-von-Neumann-era; GI - 4. Jahres-
                tagung 1974, Siefkes, G. (ed.), Lecture notes in Com-
                puter Science, Vol. 26, Springer-Verlag, 439-452, 1975

[HÄNDLER 76]   Händler, W., F. Hofmann, H.J. Schneider: A general
                purpose array with a broad spectrum of applications.
                Computer Architecture, Händler (ed.), Informatik Fach-
                berichte, Vol. 4, Springer Verlag, 311-335, 1976.

[HÄNDLER 77a]   Händler, W.: The impact of classification schemes on
                Computer Architecture; Proc. of the 1977 Int. Conf.
                Parallel Processing, J.L. Baer (ed.). IEEE. 7-15. 1977

[HÄNDLER 77b]   Händler. W.: Aspects of parallelism in computer archi-
                tecture. M. Feilmeier (ed.): Parallel Computers -
                Parallel Mathematics, North Holland, 1-8, 1977

[HÄNDLER 80]   Händler, W., H. Rohrer:
                Gedanken zu einem Rechner-Baukasten-System, Elektroni-
                sche Rechenanlagen, Vol. 22, No. 1, 3-13 (1980)

[HÄNDLER 82]   Händler, W.: Innovative computer architecture - How to
                increase parallelism but not complexity, in Parallel
                Processing Systems, 1980 Proc. Symp., Loughborough
                Univ. Technol., D.J. Evans (ed.), 1-41, Cambridge
                Univ. Press 1982

[HALLER 83]   Haller, G., R. Häuser: Entwurf und Implementierung
                eines Programms zur Steuerung eines fahrerlosen Trans-
                portsystems durch einen DIRMU-Rechner, Studienarbeiten
                am IMMD III, Universität Erlangen-Nürnberg, 1983

[HENNING 83]   Henning, W., M. Vajtersic and J. Volkert: "Matrix In-
                version Algorithm for the Parallel Computer EGPA",
                EGPA - Internal Paper (to be published)

[HOCKNEY 81]   Hockney, R.W., J.W. Eastwood: Computer Simulation
                Using Particles, McGraw-Hill, 1981

[KNEISSL 82]   Kneissl, F.: "Realisierung von Datenflußmechanismen
                auf hierarchische Mehrrechnersysteme", Arbeitsberichte
                des IMMD, Univ. Erlangen-Nürnberg, Band 15, Nr. 12,
                1982

[MAEHLE 81]   Maehle, E.: Modulare, fehlertolerante Multimikropro-
                zessorsysteme nach dem Baukastenprinzip, VDI-Berichte
                395, 91-96 (1981)
[NELDER 65]   Nelder, J.A., R. Mead: A Simplex Method for Function
                Minimization, Comp. J., Vol. 7, 308-313 (1965)
[RATHKE 83]   Rathke, M.: "SAP - Ein optimistischer Algorithmus
                für die parallele Textverarbeitung", EGPA - Internal
                Paper (to be published)
[RICHTMYER 67] Richtmyer, R.D., K.W. Morton : Difference Methods for
                Initial-Value-Problems, Interscience Pub., John Wiley
                & Sons 1967
[STRANG 73]   Strang, G., G. Fix: An Analysis of the Finite Element
                Method, Prentice Hall Inc. 1973

AN EXPERIMENTAL MODULAR MULTIPROCESSOR SYSTEM AND ITS KERNEL PROCESSING UNIT

June, 1983

by Hajime IIZUKA

Seikei University

Tokyo, Japan

1. INTRODUCTION

In parallel-processing systems, one of the most important issues in achieving high effectiveness is the coincidence of the parallel structure of both hardware and problem. Therefore, the philosophy of how problems' parallel structures are mapped into the hardware structure is very important in highly parallel computers.   Current methods can be categorized into the following types.

[*Type 1*] Parallelism depending on concurrent processing of similar jobs.

Vector or array processors are typical examples in that they carry out the same function on many data concurrently.

[*Type 2*] Parallelism depending on concurrent processing of different jobs.

Many commercial machines with multiple processors or specially-designed minicomputer complexes are examples of this type.

[*Type 3*] Distributed-function system.

These are computers consisting of many processors, each of which is dedicated to a certain system or user function.

[*Type 4*] System in which the mapping of parallelism is solely controlled by software.

This type of system has a standard yet fixed hardware parallel structure, and the accompanying software can utilize it in any way suitable for a given problem's structure. However, in this case the overhead for synchronization is usually large.

[*Type 5*] Modular multiprocessor system whose parallel structure can be tailored to each problem or class of problems.

This type of system has modular processors of a very flexible architecture with a communication scheme among them, and each particular system is structured in the most suitable way using these various features. CMU's CM* is considered a typical system of this type.

[*Type 6*] Systems in which hardware can recognize parallel structures of problems and take advantage of them in the most effective way.

So-called data-flow systems are considered to belong to this type. From the viewpoint of parallel execution, this type is very promising. However some restrictions on the program structure usually exist.

2. DESIGN PHILOSOPHY [1]

The system to be described is called ACE (Adaptive Computer Experiment). Its basic design goal is to provide the system with an ability to adapt itself to the application environment. In order to achieve this goal, the following three design features are adopted as basic approaches.

(1) Processor-Memory-Switch level modular organization and an intermodule communication system of high flexibility and generality.

(2) Extensive dynamic microprogramming capabilities. A newly-designed emulation-oriented microprocessing unit was used as the nucleus of the processor module. This microprocessor was named PULCE (Pips UniversaL Computing Element) and later was implemented on an SOS/MOS LSI chip.

(3) The structure of the standard processor module included various novel features such as dynamic microprogramming with a microcache, a data cache with 'cacheability' control, and an automatic data-length alignment.

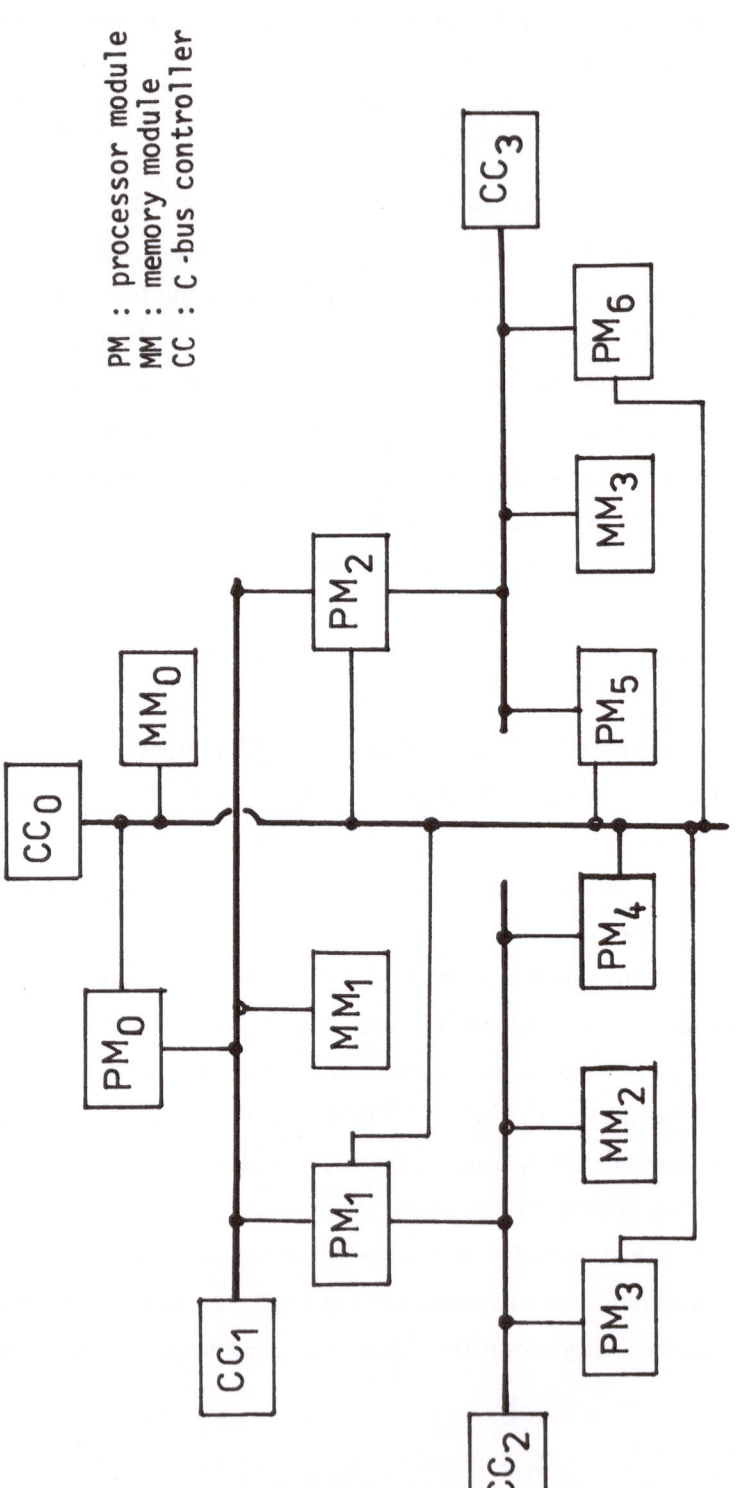

PM : processor module
MM : memory module
CC : C-bus controller

Fig. 1. A possible hierarchical configuration of the ACE system

3. FEATURES FOR INTERCONNECTIONS AMONG MODULES.

Flexibility and generality of communication among system modules are most important for modular structure computer. In ACE, all connections are achieved in one uniform way, both logically and physically, by a bus called 'C-bus'. A C-bus can be uniformly used for both processor-processor communication and processor-memory communication. Each standard processor-module has four C-bus connection ports. Taking advantage of its flexibility, the ACE system can be configured into various structures, e.g., array, hierarchy (see Fig. 1) and various hybrids.

3.1 Basic C-bus communication procedure

For the purpose of inter-module communication, the system uses global logical addresses common only on the C-bus. This Global Address (GA) may be completely independent of the local address (LA) in each module.

Every module connected to a C-bus is assigned certain ranges of the GA space (possibly dynamically) as its Recognize Address (RA) space. Whenever the requesting module puts a GA on the C-bus, each module compares it with its RA, and those which recognize the GA on the C-bus and are ready to perform the requested action respond 'Ready' and reconvert the GA to their own particular LA's, thus establishing transmission paths.

As easily seen from the above explanation, the GA may be considered as a kind of name given to the information. Accordingly, as long as the relation between the GA and the LA is fixed, the information assigned to this GA may reside in any module on the C-bus.

3.2 Broadcast communication

In the above-described communication method one-to-many communication is easily achieved, because if certain ranges of GA's are set to be recognized by all modules on a C-bus, information with these GA's are received by all the modules.

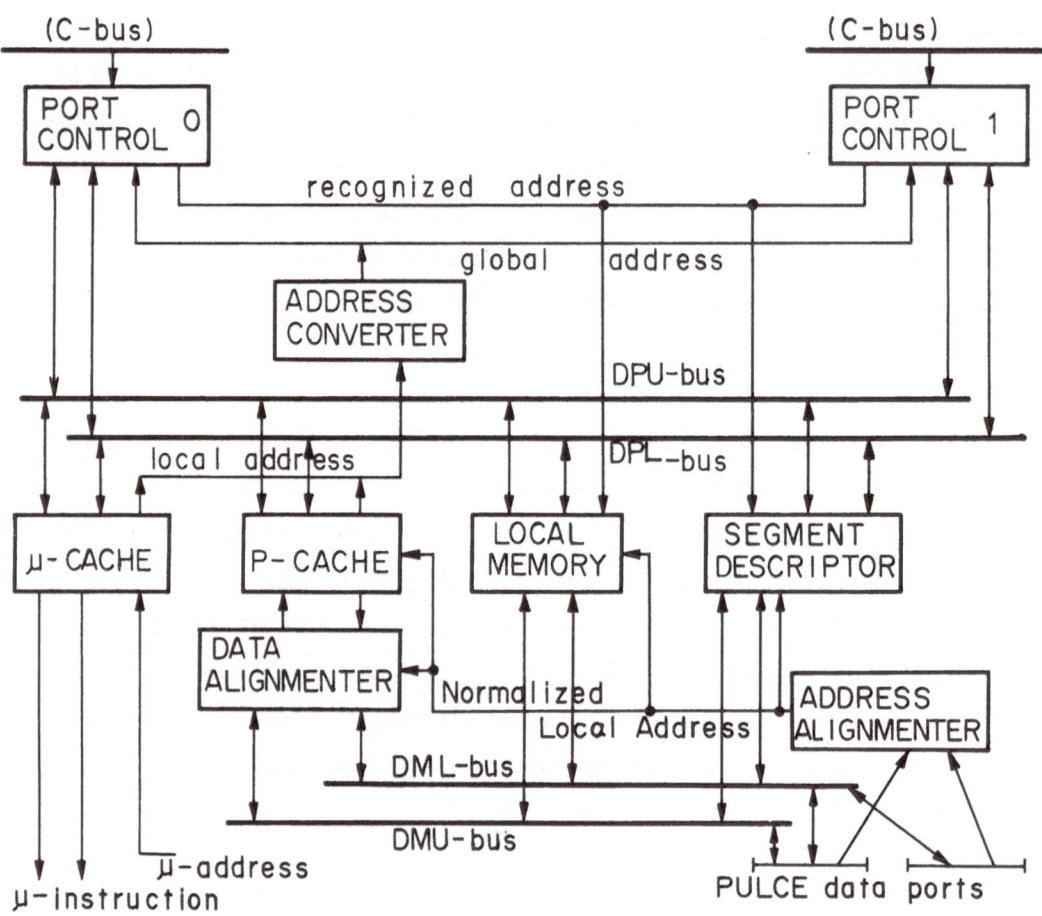

Fig. 2. Basic block diagram of the
standard processor module.

3.3 Globality

In order to achieve a hierarchical system among the modules connected to a C-bus, two-bit information which reflects the breadth of usage is defined as 'Globality' (GB). Each module compares the transmitted GB with its own port GB, and, after address recognition, participates in communication only when the received GB is equal or greater than its port GB.

3.4 Transmission on C-bus

On the C-bus, up to 16 data units are block-transferred as a compromise between transmission throu ghput and an excessive bus-holding time. To achieve flexibility a request priority was assigned to each transmission request, not to each module. Consequently, the relative weights of priorities on the C-bus become dynamically changeable.

4. STANDARD PROCESSOR MODULE

The basic processing component of ACE is called the 'standard processor module (PR-S)' which has a considerable processing power and can be easily adapted to a wide range of problems. It is constructed using a microprocessing unit with a flexible architecture (PULCE), four chunks of high-speed memory, and a considerable amount of supporting control circuits which provide PR-S with various new general-purpose emulation-oriented facilities and powerful communication capabilities.

Fig. 2 gives a basic block-diagram of PR-S. The two interfaces shown at the upper portion indicate the connection to C-buses. The major characteristics of the PR-S architecture are as follows, while details of the PULCE architecture are described in the next section.

4.1 Dynamic microprogramming

In order to give the PR-S its personality dynamically, it is provided with a dynamic microprogramming facility. A large microprogram address space (8 segments, of up to 8k steps per segment) was provided, and microprograms are stored in the same address space where data and macro-level programs are located. To speed up microinstruction access, a chunk of high-speed memory was used as a microcache.

4.2 Data cache

ACE PR-S utilizes two chunks of high speed memory as a data cache. To increase the effect of the cache even with shared information in the main memory, we decided to put a two-bit quantity called 'cacheability' in each segment descriptor to describe the nature of the cache control.

4.3 Local memory

The last chunk of a high-speed memory is used as a temporary storage and inter-module communication area.

4.4 Segmentation

The Local Address Space (LAS) used by each module and the Global Address SPACE (GAS) used for inter-module communication are separated to allow a high degree of expandability, flexibility and adaptability to the required structures. For the purpose of address translation from LAS to GAS, we have adopted a two-unit segmentation.

Since the Normalized Local Address, which is described later, is divided into a 4-bit segment number field and a 20-bit word-number field, sixteen segments are available for each PR-S. However, segment #0 is used for the local memory and registers, and is never converted into GAS.

The size of each segment is 1 to 16 multiples of either of two basic units, the larger being 64 kbits and the smaller 4 kbits. So when all the segments are of the largest size, the LAS attains a maximum size of about 2 Mbytes.

4.5 Data alignment

One of the trends of new computer applications is a increase in data types, and thus data of various sizes is to be treated in emulation. Under these circumstances, it is absolutely clear that byte and word-addressing alone are not satisfactory. Therefore, the PR-S is provided with a hardware facility for aligning the accessed data automatically.

For the data length of each segment, only 1,2,4,8,16 and 32 bits are allowed, and this is indicated in each segment descriptor. So a user can describe a data location in terms of the data unit of the corresponding segment. The address alignment hardware of the PR-S converts the user-supplied LA to a length-independent bit-address, called Normalized Local Address (NLA).

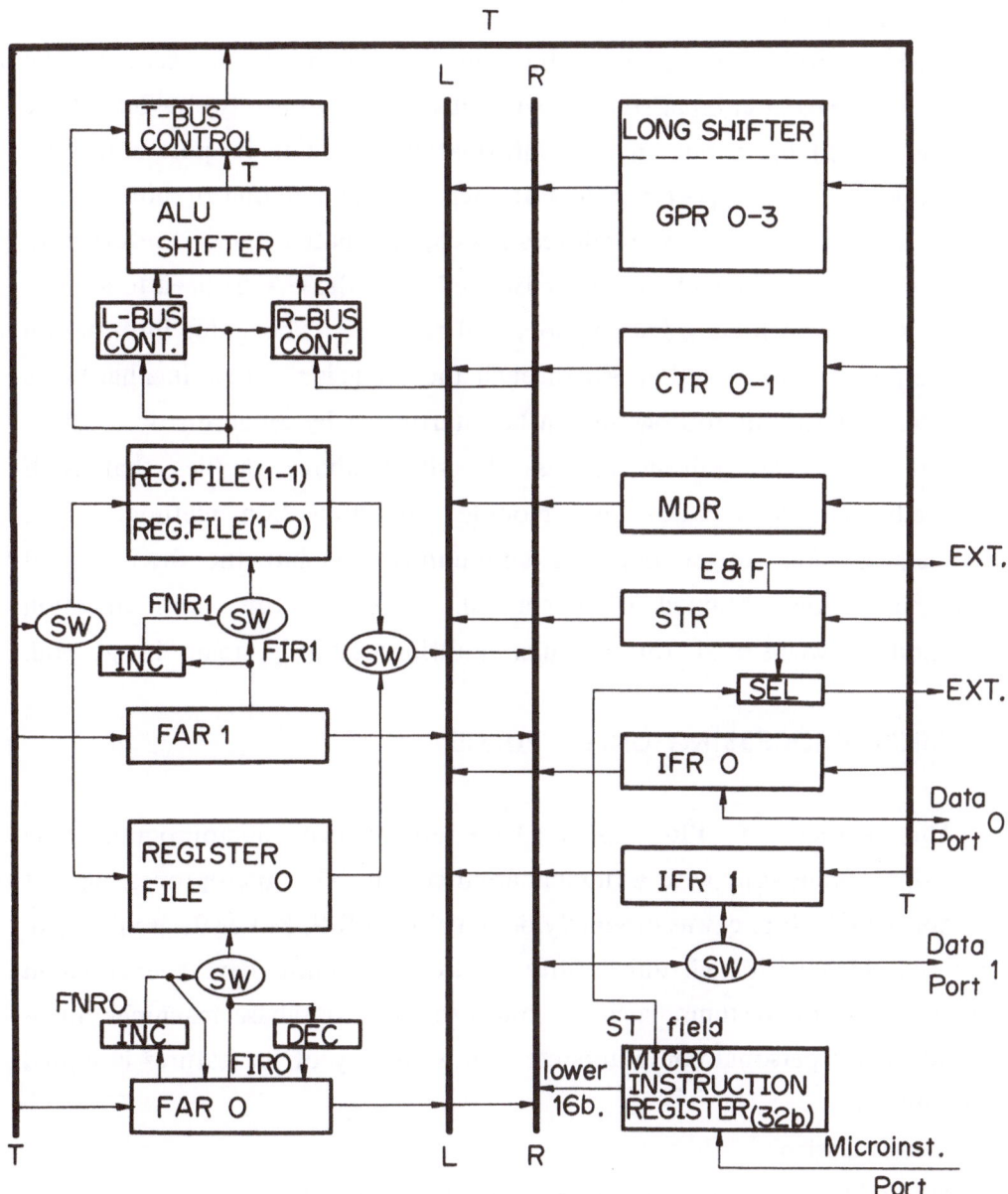

Fig. 3. Design of the PULCE microprocessing unit.

GPR: General Purpose register, CTR: Counter, MDR: Mode register
STR: Status register, IFR: Interface register, FAR: File address
register, FIR: File indirect register, FNR: File next register,
INC: Incrementer, DEC: Decrementer, SEL: Bit selector, SW: Switch.

4.6 Port control

PR-S has four C-bus ports of the same specifications. To recognize the GA on a C-bus, a comparison is made between the upper twelve bits of the GA issued on the C-bus and the information stored in its port registers. There are two such port registers at each port and if one or both of the active registers recognizes an address match, the match signal is returned to the C-bus controller and the lower twelve bits of the GA are used to address the PR-S's local high speed memory and registers. The status information concerning an attempted access through the C-bus is recorded in a hardware register and the microprogram can be interrupted by an access occurrence.

The address-recognition process described above implies that ACE basically uses a mail-box inter-module communication method. This process requires longer time for communication, but the flexibility of communication obtained is substantial. Moreover, this communication method is well-suited for modular multiprocessor systems in general.

5. MICROPROCESSING UNIT — PULCE

The kernel of PR-S is a high-performance microprogrammable microprocessing unit with a flexible architecture. This microprocessing unit is called PULCE and was originally designed for PR-S. But as its architecture was provided with high universality, PULCE LSI chips have been used in many kinds of systems, such as multiprocessor database machines, high-performance personal computers etc. The summary of its features is shown in Table 1.

5.1 Basic Design

The following are the basic design considerations.

(1) Semiconductor technology nMOS/SOS was used as an implementation technology to achieve high performance.

(2) Only arithmetic function and registers were included in the PULCE LSI, and a sequence-control function was tailored to a specific application outside LSI. This was due to flexibility and limitation in the number of gates on a chip at the time this was designed.

Device type	n MOS/SOS
Chip size	8.85x6.66mm
Gates in a chip	7000
Transistors in a chip	20000
Package	80-pin flat package with cooling fins.
Power supply	5V
Machine cycle	200ns
Power dissipation	1.5W
Operating temperature	0°C-50°C
Data width	16bits
Microinstruction	32bits supplied from outside
Registers (General purpose) (Mask) (Dedicated)	44 29 7 (16bits) 6 (4bits) 2
Shifter (Single word) (2,3,4 words)	0-15bits 1bit
Decimal operation	add/sub (1 digit)
Stack	Hardware support
Multiply/divide	Special hardware support instruction

SUMMARY OF THE PERFORMANCE OF PULCE

Table 1

(3) 16 bits was chosen as the basic word length. But 32- bit general-purpose interface and some architectural features for 32-bit processing were provided.

(4) PR-S is intended to be used as a universal host processor. Thus, PULCE was provided with an emulation- oriented architecture. For this purpose, features such as data field masking, indirect access to register files and a few operation-mode-control bits were provided.

(5) To increase the speed of stack operations, PULCE was equipped with a special hardware support that always keeps the upper portion of a stack in internal registers.

(6) For the best performance of PULCE, a flexible control of hardware through easy microprogramming is very important. Therefore microprogram control by 32-bit vertical microinstructions with a horizontal flavor has been adopted. In addition, the organization of the microinstruction repertoire and internal structure was made as regular as possible. Therefore, users of PULCE have the flexibility to control hardware details through relatively easy microprogramming. For a more detailed description, refer to the papers[2,3].

6. CONCLUSION

We have seen an architecture of an experimental modular multiprocessor system and a microprogrammable microprocessing unit with a flexible architecture. A prototype of ACE was constructed at the Electrotechnical Laboratory, which consisted of 3 processor modules, 2 memory modules with 16k words of 32 bits/word each, a commercial minicomputer as I/O processor, and a special synchronizing module connected by 2 C-buses; and some software, including concurrent Pascal machine and an operating system which supported user-microprogramming, was also developed and evaluated. The first phase of the study is completed, and a new study, essentially along the same lines, is in the planning stages, but the new study will accomodate recent VLSI developments.

REFERENCES

(1) H.Iizuka et al. "ACE — A new modular computer architecture", Proc. 2nd USA-Japan Computer Conference, pp.36-41(1975)

(2) H.Iizuka et al. "Development of a high-performance universal computing element — PULCE", Proc.NCC, pp.1255- 1264 (1978)

(3) H.Iizuka "Design and implementation of a microprocessing unit with flexible architecture", pp.22-38, Computer Science & Technologies 1982, OHM-North Holland

Springer Series in

Information Sciences

Editors: **K.-s.Fu, T.S.Huang, M.R.Schroeder**

Springer-Verlag
Berlin
Heidelberg
New York
Tokyo

Lecture Notes in Physics

Selected Issues from
Lecture Notes in Mathematics